Multivariable Predictive Control

Multivariable Predictive Control

Applications in Industry

Sandip Kumar Lahiri

Registered Offices
John Wiley & Sons, Inc., 111 River Street, Hoboken, NJ 07030, USA
John Wiley & Sons Ltd, The Atrium, Southern Gate, Chichester, West Sussex, PO19 8SQ, UK

Editorial Office
111 River Street, Hoboken, NJ 07030, USA
The Atrium, Southern Gate, Chichester, West Sussex, PO19 8SQ, UK

For details of our global editorial offices, customer services, and more information about Wiley products visit us at www.wiley.com.

Wiley also publishes its books in a variety of electronic formats and by print-on-demand. Some content that appears in standard print versions of this book may not be available in other formats.

Library of Congress Cataloging-in-Publication Data

Names: Lahiri, Sandip Kumar, 1970– author.
Title: Multivariable predictive control : applications in industry / Sandip Kumar Lahiri,
    Supra International Private Ltd, Vadodara, India.
Description: First edition. | Hoboken, NJ, USA : Wiley, 2017. | Includes bibliographical
    references and index. | Description based on print version record and CIP data
    provided by publisher; resource not viewed.
Identifiers: LCCN 2017010553 (print) | LCCN 2017012540 (ebook) | ISBN 9781119243519 (pdf) |
    ISBN 9781119243595 (epub) | ISBN 9781119243601 (cloth)
Subjects: LCSH: Predictive control. | Multivariate analysis.
Classification: LCC TJ217.6 (ebook) | LCC TJ217.6 .L34 2017 (print) | DDC 629.8–dc23
LC record available at https://lccn.loc.gov/2017010553

Cover design by Wiley
Front Cover Image: FotoBug11/Shutterstock
Back Cover Image: Paulo Vilela/Shutterstock

Set in 10/12pt Warnock by SPi Global, Pondicherry, India
Printed and bound in Malaysia by Vivar Printing Sdn Bhd

10   9   8   7   6   5   4   3   2   1

To my parents, wife Jinia and two lovely children Suchetona and Srijon

Contents

Figure List

Table List

Preface

In chemical process industries, there is an ongoing need to reduce cost of production and increase profit margin. Due to cut-throat competition on a global level, the major chemical industries are now competing to optimize raw material and utility consumption, to reduce waste, to reduce emission, and to minimize pollution. Multivariable model predictive control (MPC) is considered as an excellent tool to achieve those goals. The benefit of implementing MPC are many. MPC optimizes the plant operation on a continuous basis, reduces waste and utility consumption, minimizes raw material consumption, and maximizes production. Due to these benefits, all major chemical industries, petrochemical industries, and oil refineries throughout the globe are implementing MPC in their plants.

However, there are no dedicated books available to discuss the basic concepts of MPC, provide practical guidelines, and explain industrial application procedures.

The main idea of writing this book is to fill this gap with the following people in mind: managers, process engineers, control engineers, operators working in the process industries, and chemical engineering students who want to pursue process control career.

MPC is normally implemented by an external MPC consultant company or experts such as AspenTech, Honeywell, and Shell. The practicing process engineers or process control engineers working in the plant normally have much less exposure or knowledge to implement MPC. The available books in market on MPC don't cover the practical aspects to implement commercial MPC software.

The available books on MPC emphasize unnecessary theoretical details, which are normally not required by the practicing engineers, and those theories have very little relevance for commercial implementation of MPC software. This book discusses the practical aspects of MPC implementation and maintenance. The consultants or experts coming from MPC vendor companies normally implement MPC, hand over the technology to client plant, and then leave. After they leave, the responsibility goes to plant process engineers and control engineers to keep the MPC software running, derive maximum benefit from it, and sustain those benefits by proactive maintenance. So plant engineers need to have a thorough understanding about the different features of MPC software and key implementation steps. Often, due to unavailability of literature on this subject, plant engineers lack the knowledge and understanding of MPC.

The book is intended to build an overall understanding of MPC implementation and how to derive maximum benefit from MPC. It covers everything that a practicing process engineer or process control engineer needs to know to build an effective MPC application. Practical considerations of MPC implementation are emphasized over unnecessary theoretical details. The book covers a wide range of subjects of MPC applications, starting from an initial functional design stage to final implementation stage. Readers will also get enlightened as to why many MPC applications fail in industries across the globe. The root causes of this failure are discussed in detail so that readers of the book can safeguard and take preventive and corrective action beforehand to avoid MPC failure.

As this book covers a wide range of topics, the materials are organized in such a way that helps the reader to locate the relevant chapters quickly, to be able to understand them readily, and to apply them in the right context. The book is organized in the following way.

Overview of Contents

Chapter 1 gives an overview of the importance of multivariable predictive control (MPC) in chemical process industries in the context of today's competitive business environment. The benefits of implementing MPC over normal Proportional-Integral-Derivative (PID)-type regulatory control and how MPC brings this benefit in real commercial chemical plant are explained in detail here. A brief description of MPC working principle is also discussed. The purpose of process control in chemical process industries (CPIs) is to ensure safety, maintain product quality and operational constraints while trying to maximize economic benefit. Traditionally, PID controllers are used in CPIs. However, PID controllers are not efficient to handle multivariable processes with significant interactions. Multivariable model predictive optimizing controller understands these process interactions and makes multiple small moves with the help of its model predictive capability. By doing this, it slowly brings the process to the most economic operating zone while maintaining all the process parameters within their limits. MPC acts as a supervisory controller above base-level PID control and is situated at the middle of a multilevel control hierarchy. The relevance of multivariable predictive control (MPC) in chemical process industry in today's business environment is very high while industries are struggling to reduce operating cost, maximize profit margin, and reduce waste. MPC stabilizes the process by utilizing its model predictive capability and thus allows the operation near to constraints. MPC is applied in oil refinery, petrochemical, fertilizer, and chemical plants across the globe and they bring huge amount of profit. The chapter ends with practical examples of MPC implementations in various process industries starting from petrochemicals, petroleum refinery to fertilizer and many other chemical plants.

Chapter 2 deals with theoretical foundation of MPC. Different variables and commonly used terms in MPC are introduced in the chapter. Different features of MPC controller are explained in detail. A simple algorithm explains the reader the underlying calculation steps of MPC technology. Simplified dynamic control strategy of MPC controllers are discussed in detail to develop an understanding of how it works. One of the major features of MPC is its future prediction and constraint handling capability. The theoretical background of these two main features is explained in detail with examples.

Model predictive controllers (MPCs) have many features. They are multivariable controllers with model-based predictive capability. They continuously optimize the process by rigorously planning and executing small movement in manipulated variables (MVs). As simple architecture, they have data collection module, control variable (CV) prediction module, steady-state optimization module, and dynamic optimization module. The process starts with reading current value of controlled variable and MV, and using its internal process model it predicts the future value of controlled variable. In every execution, it reconciles this prediction value with actual process measurements to compensate for model inaccuracies. Also, it calculates the size of the control process (i.e., number of MV and CV available for control purpose) in every execution and sees whether any change in size is made by panel operator. Its steady-state optimization module then calculates the optimum targets for CV and MV and feeds this information to dynamic module to plan detail MV movement to achieve those targets. Depending on various tuning parameters and MV–CV limits, dynamic module initially plans for a series of MV movements so that those targets can be achieved and process can be brought to the most economic optimum zone. The first step of MV movement is actually implemented through PID controllers and all other moves are discarded. In next execution, again all the calculations are repeated.

The chapter explains all of the aforementioned features in a simple way.

Historical developments of different MPC technology are described in detail in Chapter 3. First-generation MPC was developed in 1970s. Over the years, MPC technology went through various modifications and additions of different features and reached currently as fifth generations MPC technology. The genesis of these developments over the years, the need, and innovations at different generations are discussed in this chapter. MPC control algorithm is developed over the years starting from 1970s. The initial IDCOM, an acronym for Identification and Command and Dynamic matrix control (DMC) algorithms represent the first generation of MPC technology (1970–1980); they had an enormous impact on industrial process control and served to define the industrial MPC paradigm. Engineers at Shell Oil continued to develop the MPC algorithm and addressed the weakness of first-generation algorithm by injecting quadratic program (QP) in DMC algorithm. The QDMC algorithm can be regarded as representing a second generation (1980–1985) of MPC technology, comprising algorithms that provide a systematic way of implementing input and output constraints. However, after initial phase, MPC technology slowly started to get huge profit and gain wider acceptance during the 1990s. The Identification and Command, modified version (IDCOM-M), Hierarchical constraint control (HIECON), Single Multivariable Control Architecture (SMCA), and Shell Multivariable Optimizing Controller (SMOC) algorithms represent a third generation of MPC technology (1985–1990); others include the predictive control technology (PCT) algorithm sold by Profimatics, and the RMPC algorithm sold by Honeywell. In the era of 1990–2000, increased competition and the mergers of several MPC vendors have led to significant changes in the industrial MPC landscape. Major MPC companies started acquisition and wanted to dominate the market. AspenTech and Honeywell got out as the winners of this phase and represent fourth-generation MPC (1990–2000). Today, we are witnessing a further technology development that is not so much focused on improving the algorithms, but to improve the development steps. This represents fifth-generation algorithm (2000–2015). The focus is put to make those steps smoother, faster, and easier, for both the developer and the client, and to do as much as possible remotely. The chapter enlightens readers on all of the aforementioned areas.

Implementing MPC in chemical plants is itself a project and involves lot of steps. Chapter 4 gives an overview about the various stages of MPC implementation starting from an assessment of existing regulatory control, functional design of MPC, model building and final MPC implementation stages. It starts with preliminary cost–benefit analysis to evaluate approximate payback period. Assessment of base control loop and strengthening it is a basic requirement to build a solid foundation upon which MPC works. In functional design step, a list of controlled and MVs are identified. Plant step test is carried out to collect dynamic data of CV for a step change in MV. These step test data are utilized to build models in model building stage. Potential soft sensors are made where online analyzers are either not available or very costly. The suitability of developed model for control purpose is checked in off-line simulation mode. After that, controller is commissioned in actual plant and online tuning is done to achieve the desired controller action. As a last step, performance monitoring and benefit assessment of installed MPC controller is done. An essential part of each step is to train the plant operators and engineer regarding different features of MPC and how to operate the installed MPC application. The chapter also explains the steps involved in MPC projects with vendor.

Normally, the implementation of MPC involves cost that includes MPC software, hardware cost, and MPC vendor cost. Client plants who want to implement MPC always want to know about the benefit or payback period of MPC implementations before they decide to go for MPC implementation. Chapter 5 describes cost–benefit analysis procedures before MPC implementation.

Preliminary cost–benefit analysis is usually carried out before starting MPC project. The purpose is to estimate the actual benefit after MPC implementation. A scouting study of process analysis and economic opportunity analysis is done to know the potential areas where MPC can bring profit. By its model-based predictive capability MPC stabilizes the process and reduces variability of key process parameters. This reduction of variability enables operators to shift the set point closer to the constraints. Operation closer to constraints translated into more profit. By statistical analysis, this increase of profit due to MPC implementation is calculated. Finally, a scientific cost–benefit analysis is done to evaluate the payback period. The results of the cost–benefit analysis help the plant management to take economic decision to implement MPC in plant. An example with practical case study is also given to explain the cost–benefit analysis procedure.

Chapter 6 explains the procedure to assess the health of regulatory base control layer of plant. MPC cannot work efficiently if base control layer or regulatory control layer is weak. Hence, strengthening base control layer is an important prerequisite to build the good foundations of MPC. Over the years, process industries technical community realizes the importance of monitoring the base control loop performance. The benefits gained from detecting the weakly performed control loop and subsequently improving their performance are huge. Assessment of regulatory base control layer in plant starts with understanding different common failure mode of valves, sensor, controller, and so on. Control valves may malfunction due to hysteresis, stickiness, and improper valve sizing. Sensors exhibit different problems such as noisy indication, improper calibration, and overfiltration, to name a few. Controllers commonly have tuning problems. Sometimes, process also has problems such as variable gains and too much interaction. Due to a large number of control loops present in any moderate-sized process industries, manual evaluation of each control loop performance is not feasible. Online systematic performance monitoring of control loops through various key performance indices (KPIs) and matrices is the need of the hour. This gives rise to a new technology/ software called control performance monitoring/assessment (CPM/CPA). Performance KPIs are generated and monitored online, and they are grouped as follows: traditional KPIs, statistical-based metrics, business/operational metrics, and advanced indices. The chapter ends with giving a short exposure of controller tuning for PID controllers.

Functional design is the most important step in MPC project. Functional design is the proper planning and design of MPC controller to achieve operational and economic objective of the plant. There is no standard procedure to be followed to do a functional design. It depends on expertise and experience of MPC vendor or control engineer, plant operating people, and plant process engineering people.

Chapter 7 explains in detail about various aspects and practical considerations of functional designs of MPC controller in actual commercial plants. This step starts with understanding of process opportunity and process constraints. Process controls objective, controller scope, and identification of CV–MV–DV list is done in this step. Exploring the potential optimization opportunity is a key job in functional design stage. Identification of any scope to implement the inferential calculations or soft quality estimators is also done in this stage. Conceptualization of economic objective of controller and form of linear program (LP) and quadratic program (QP) objective function is finalized in this step.

Functional design of MPC controller started with the identification of controlled and MV and subsequent planning for MPC model structures. Practical considerations to identify process and equipment constraints are also discussed to help the reader formulate a robust, safe, and reliable MPC model. Good step test data is of paramount importance in MPC model building and its overall functioning. How to ideally perform step test in actual shop floor of the plant and do's and don'ts of step test are discussed in detail. The chapter also briefly explains the requirement of soft sensor building.

Chapter 8 deals with preliminary process step and step test. Step test is considered as one of the major steps in MPC project. In step testing, step change in MVs is given and the impact of it on CVs with time is collected through step test data. These data are used to build process model. Both open-loop and closed-loop test are practiced in industry, and both methods have their own advantages and limitations. As the MPC models are data-driven empirical models and those data are generated in step test, it is very important to carry out this test with all precautions. The quality of developed model will be as good or as bad as step test data. Hence, it is important to know all do's and don'ts of step testing method. To reduce the unnecessary problems in step test, a preliminary process test or pre-stepping is done before step test. The purpose is to identify all the possible bad actors of step test and rectify them beforehand. The chapter explains various do's and don'ts in step test.

Chapter 9 describes in detail about the various model building procedures available in commercial software. Process models are dynamic MV–CV relationship generated from step test data. In model building step, a suitable model structure with proper order is first identified. Later on, model coefficients are evaluated from step test data by statistical fitting operation. Various data cleaning methods and outliers detection are discussed. Basic steps of process identifications start with execution of step test and collection of data, pre-processing and cleaning of data, selection of model structure and order, and determination of model parameters. There are a lot of predefined dynamic model structures available in the library of commercial identification software. Knowing those structures and their key strength and weakness and finally identifying a suitable structure to accurately model the step data is the key of system identification step.

Theoretical background of various available models and their implication in MPC is explained in detail in the chapter. One of the major requirements for robustness of MPC model is to validate the developed data-driven MPC model from practical process knowledge so that the model captures the underlying physics of the process. This important aspect is discussed in detail to give the reader a flavor regarding efficient and robust model building. Practical considerations to prioritize MVs to control particular CV in multivariable environment are discussed in detail so that user can maximize the economic benefit after MPC implementation.

An inferential or soft sensor is a mathematical relation that calculates or predicts a controlled property using other available process data. When it is very difficult or costly to measure an important parameter online, such as distillation tower top product impurity, soft sensors are used to predict that inferential property from other easy measurable parameters such as top temperature and pressure. Sometimes, soft sensors are used as backup of an existing analyzer to reduce or eliminate dead time, both from the process and the analyzer cycle.

Chapter 10 is dedicated for soft sensors available in various process industries. What are soft sensors and how to make them is the main idea of the chapter. Various commonly used algorithms to build fast principle-based and black-box-based soft sensors models are discussed in detail. Why some soft sensors fail in industry and precautions needed to make successful robust soft sensors are described in the chapter.

Usually, four types of soft sensors are used in industry, namely, first principle–based soft sensor, data-driven soft sensors, gray model–based soft sensors, and hybrid model–based soft sensors. There are many methods to develop industrial soft sensors and usually they include the following steps: data collection and data inspection, data preprocessing and data conditioning, selection of relevant input–output variables, aligning data, model selection, training and validation, analyzing dynamics, and finally deployment and maintenance.

Due to the difficulties in developing first principle–based soft sensors, data-driven soft sensors are very popular in industry. Major data-driven methods for soft sensing which dominates the industry and discussed in this chapter are principle component analysis, partial least squares, artificial neural networks, neuro-fuzzy systems, and support vector machines.

After development of process model, it is important to know how the developed controller will perform in online mode before its actual deployment in real plant. Off-line simulation refers to running the controller in a separate off-line PC to see the MV–CV dynamic responses of the process. One major task of off-line simulation is to set the different tuning parameters of the controller. The purpose is to perform off-line tuning and other corrections as much as possible so that the application runs effortlessly in actual plant at real time.

Chapter 11 is dedicated to off-line simulation of MPC model—an important prerequisite step for the MPC online implementation. How to set up off-line simulation in MPC software and how to derive maximum benefit from them is the main focus of the chapter. Constraint handling capability of developed MPC model can be assessed in off-line simulation. How to learn and modify the MPC model structure and tuning parameters from off-line simulation response is explained in detail in the chapter.

Off-line tuning involves setting proper priority for CV and MVs, CV give up priority in case of infeasible solution, setting up optimizer speed and different coefficient of LP and QP objective function. Before starting off-line simulations, it is important to understand the concept of different tuning parameters available in MPC software package. It is also important to understand how MPC works in a dynamic environment and how different tuning parameters can impact its performance. These are explained in detail in the chapter. Usually, there are three major categories of tuning parameters, namely, tuning parameters for CVs, tuning parameters for MVs, and tuning parameters for optimizer. Various simulation tests can be planned, configured, and run in off-line simulator to assess the different features and functionality of the developed controller in off-line mode. Changes in different tuning parameters are done on trial-and-error basis until a satisfactory dynamic performance of MPC controller is achieved.

Online deployment of MPC application in real plant means connecting the MPC controller online with the plant Distributed control system (DCS) and allowing it to take control of the plant.

Chapter 12 is dedicated to the most important steps in MPC implementation—online deployment in real plant. Various stages of real-time deployment of MPC software and precautions to be taken in open- and closed-loop deployment are described in detail. Unless these precautions are taken, MPC software may lead to bumpy control of processes and shutdown of the plant in worst case. Different vendors have different methodology to commission the controller. However, the basic steps remain the same and are as follows: setting up the controller configuration and final review of the model, building the controller, load operator station on PC near the panel operator, taking MPC controller in line with prediction mode, putting the MPC controller in closed loop with one CV at a time, observation of MPC controller performance, putting optimizer in line and observation of optimizer performance, evaluation of overall controller performance, and online tuning and troubleshooting. Monitoring of MPC model performance after deployment and understanding the weakness of the developed model (if any) is key to make robust MPC application. Readers can gain insights of these features in the chapter. It is important to understand the purpose and details of implementations of the aforementioned step to avoid any malfunctioning of controller during commissioning phase. Care should be taken such that any mistake during commissioning does not lead to plant shutdown or plant upset. After the controller commissioning, proper documentation, and training of operators, engineers on online platform was usually done. Later on, some adjustments in different MV–CV limits and controller tuning parameters are done periodically to sustain the benefit of MPC controller.

Chapter 13 is dedicated to an important aspect—MPC controller online tuning.

Online controller tuning means changing of various tuning parameters of controllers online so that an optimal and expected performance is achieved by MPC controller in actual plant environment. If performance of MPC controller is not at par, it is recommended to investigate the root cause and troubleshoot the problem rather than jumping to tune the controller. How to systematically investigate and troubleshoot the problems of MPC controller is discussed in the chapter. As MPC is a multivariable controller; any change of one tuning parameters will affect many CVs and other MVs movement. It is important to understand the impacts of various tuning parameters on dynamic performance of controller during online tuning. There is always balance and compromise in online tuning. How to get that delicate balance is key to controller tuning. With proper knowledge, tuning parameters are modified by trial and error in online controller until an acceptable optimal dynamic performance is achieved.

Unlike PID controller tuning, MPC controller tuning involves many parameters and requires deep knowledge of MPC functioning and its overall impact on process. After reading the chapter, the reader can understand various free parameters available for tuning and how to tune MPC controller to make it robust and efficient.

The chapter describes various limits and constraints applied on MV and CV in commercial MPC packages. The idea behind putting operator limit, steady-state limit, engineering limit, and so on is discussed. How these limits impact the overall MPC performance is explained. Practical considerations to set these limits to gain maximum benefit are also discussed.

Chapter 14 enlightened the reader why some MPC application fails in industries. There are several instances all over the world that MPC brings huge profit just for 1–2 years of its implementation and then profit starts decreasing. In extreme case, some oil refineries reported that MPC application does not generate any additional benefits even after 4–5 years of implementation over the simple regularity control. Different root causes of MPC failure in industry are explained to give the reader an idea about what can go wrong. User can gain insights about how to safeguard MPC performance deterioration in the long run. The chapter ends with the various unsuccessful and failure case study of MPC application in various industries across the globe.

There are two modes of failure, namely, failure to build efficient MPC application when it was first build and gradual deterioration of MPC performance post implementation. Reasons such as capability of technology to capture benefit, expertise of implementation team, and reliability of Advance process control (APC) project methodology are responsible for the failure after it was first build. Contributing failure factors of post implementation of MPC application are attributed to the following: lack of performance monitoring of MPC application, unresolved basic control problems, poor tuning and degraded model quality, and significant process modifications and enhancement. Not only technical factor but also nontechnical failure factors are responsible for MPC gradual performance degradation. Lack of properly trained personnel, lack of standards and guidelines to MPC support personnel, lack of organizational collaboration and alignment, and poor management of control system are some of the nontechnical failure factors that need proper attention. There are three solutions, namely, technical solutions, management solutions, and outsourcing solutions to deal with MPC performance deterioration. Development of online performance monitoring of APC applications, improvement of base control layer, training of MPC engineer and console operators, development of Corporate MPC standards and guidelines, central engineering support organization for MPC, and outsourcing the solutions to MPC vendors are the major strategies to sustain MPC benefits over the years. The chapter describes all these in detail.

Chapter 15 describes the final steps of MPC implementation—its actual performance assessment after deployment in real plant. Reader can gain insight about the controller performance and the optimizer performance and how to quantify them in real monetary terms. What to monitor to assess the performance in long term is also discussed in the chapter.

Performance assessment after MPC implementation will give a true picture of how much profits are achieved by a particular application as compared with initial study before implementation. A periodical performance review will also provide an idea of how much of initial benefits are preserved over time and how much money is lost by not getting the full potential performance. This will help to justify periodical maintenance or overhaul of MPC application. Performance of model predictive control application (MPCA) can be evaluated using the following four categories: control performance (whether it is able to control all its key parameters within their desired range or not), optimization performance (whether it is able to run the plant at its limit or constraints to maximize economic benefit), economic performance (how much MPCA increase profit before and after implementation in money terms), and nontangible performance (how much operator time it saves to monitor DCS). Usually, different KPIs are created and monitored periodically for each of the aforementioned cases to determine whether performance is deteriorating over time. It is important to understand the definitions and underlying calculations of these KPIs along with their implications to safeguard the MPC performance deterioration over time. Some of the major KPIs are service factor, KPIs for financial performance, KPI for standard deviation of key process variable, KPI for constraint activity, KPI for constraint violation, KPI for inferential model monitoring, model quality, limit change frequencies for CV/MVs, active MV limit, KPIs for long-term performance monitoring of MPC, and so on. Once performance deterioration is detected by these higher level KPIs, then some low-level detail KPIs are dig down to know the actual problems and troubleshoot them. KPIs to troubleshoot poor performance of multivariable controls include KPIs for poor performance of the controller itself, KPIs to troubleshoot cycling, KPI for oscillation detection, KPIs for regulatory control issues, KPIs for measuring operator actions, and KPIs for measuring process changes and disturbances. How to create these KPIs and what is their significance and implications are discussed in detail in the chapter.

Chapter 16 gives an idea about the various available commercial MPC vendors and their applications. A comparative study of various MPC software available in market such as Aspen, Honeywell, and SMOC has been made. They all have different implementation strategies and different unique features and relative strengths and weakness. All these are discussed in detail in the chapter. Readers can get a flavor of commercial MPC applications in chemical process industries across the globe. MPC is a matured but constantly evolving technology. Although there is no breakthrough development in core MPC algorithm in the past five years, the commercial MPC vendors comes up with more software packages, which helps to implement and monitor MPC technology in shop floor. These MPC vendors are now offered a full range of software package that comprises some basic modules such as data collection module, MPC online controller, operator/engineer station, system identification module, PC-based off-line simulation package, control performance monitoring and diagnostics software, and soft sensor module (also called quality estimator module). What these different modules intended to do and various common features of these modules in commercial MPC software are discussed in detail in the chapter. The chapter also describes development history and features of three major MPC players, namely, Aspen Tech DMC-plus, Shell Global Solutions SMOC, and Honeywell's RMPCT. The discussion of the commercial MPC vendor and their software revolves around the following: a brief history of the development of each MPC technology, product offerings of each vendor with some of their uncommon features, and distinctive feature of their respective technology with current advancement.

The main feature of the book that differentiates it from other MPC books on the markets is its practical content, which helps readers to understand all steps of MPC implementation in actual commercial plants. The book describes in detail initial cost–benefit analysis of MPC project, MPC software implementation steps, practical considerations to implement MPC application, the steps to take after implementation, monitoring of MPC software, and evaluating its post-performance.

Key features of the book are summarized as follows:

- Readers can develop a thorough understanding of steps for building a commercial MPC application in a real plant. All the practical considerations to build and deploy an MPC model in commercial running plants are the essence of the book.
- Chapter 5 describes cost–benefit analysis procedures before MPC implementation.
- The stages of commercial MPC implementation, starting from an assessment of existing regulatory control, functional design of MPC, model building, and final MPC implementation stages are described in detail.
- The various aspects and practical considerations of functional designs of MPC controller in actual commercial plants are discussed.
- Soft sensors are discussed in detail in Chapter 10. Commonly used algorithms to build first principle-based and black-box-based soft sensors models are explained.
- How to learn and modify the MPC model structure and tuning parameters from off-line simulation response is explained in detail.
- Chapter 12 is dedicated to the most important steps in MPC implementation— online deployment in real plants. Various stages of real-time deployment of MPC software and precautions to be taken in open- and closed-loop deployment are described in detail.
- Monitoring of MPC model performance after deployment and understanding the weakness of the developed model (if any) is key to make robust MPC application. Readers can gain insights of these features in Chapter 15.
- The book enlightens the reader as to why some MPC applications fail in industries. Different root causes of MPC failure in industry is explained to give the reader an idea about what can go wrong. Users can gain insights about how to safeguard MPC performance deterioration in the long run.
- Chapter 16 closes out the book with a discussion of commercial MPC software applications with their distinctive features.

It is my sincere hope that readers will find the methods and techniques discussed in the book useful for understanding, functional design of MPC application, online and off-line tuning and post monitoring of MPC. This will help the readers to build and implement effective MPC application in industry and get maximum benefit from it.

Clearly, it was not a small effort to write the book, but the absence of such MPC book in market and its requirement in large number of process industries spurred me to writing. I would like to thank Mr. Farid Khan of Reliance Industry Ltd for teaching me the basics of MPC and exposing me to the practical field of MPC.

Finally, I am truly grateful to my family: my wife Jinia and my children Suchetona and Srijon for their understanding, support, and generosity of spirit in tolerating my absence during the writing of this book.

May 2017 *Sandip Kumar Lahiri*

1

Introduction of Model Predictive Control

1.1 Purpose of Process Control in Chemical Process Industries (CPI)

Any industrial process, especially oil refineries and chemical plants, must satisfy several requirements imposed by its design and by technical, economic, and social conditions in the presence of ever-changing external influences (disturbances). Among such requirements, the most important ones are as follows:

- *Safety:* This is the most important requirement for the well-being of the people in and around the plant and for its continued contribution to economic development. Thus, the operating pressures, temperatures, concentrations of chemicals, and so on should always be within allowable limits.
- *Product quality/quantity:* A plant should produce the desired quantity and quality of the final products.
- *Environmental regulations:* Various international and state laws may limit the range of specifications of the effluents from a plant (e.g., for ecological reasons).
- *Operational constraints:* The various types of equipment used in a chemical plant have constraints (limits) inherent to their operation. Such constraints should be satisfied throughout the operation of the plant (e.g., tanks should not overflow or go dry).
- *Economics:* The operation of the plant should be as economical as possible in its utilization of raw materials, energy, and human labor.
- *Reliability:* The operation of the plant should be as reliable as possible to ensure that the plant is always available to make products.

These requirements dictate the need for continuous monitoring and control of the operation of a process plant to ensure that operational objectives are met. This is accomplished through an arrangement of instrumentation and control equipment (measuring devices, valves, controllers, computers) and human intervention (plant designers, plant operators), which together constitute the control system.

There are three general classes of requirement that a control system is called on to satisfy:

1) Suppression of disturbances
2) Ensuring the stability of the process
3) Optimizing the performance of the process

Multivariable Predictive Control: Applications in Industry, First Edition. Sandip Kumar Lahiri.
© 2017 John Wiley & Sons Ltd. Published 2017 by John Wiley & Sons Ltd.

Traditionally, PID controllers are used in CPI to perform these tasks. PID regulatory controllers efficiently ensure stability of the process and suppression of disturbance. However, due to the multivariable nature of the process and complex interactions between process parameters, PID controllers cannot make a coordinated control move to optimize the process performance. Here lies the need of model predictive control.

1.2 Shortcomings of Simple Regulatory PID Control

PID control forms the backbone of control systems and is found in most CPI. PID control has acted very efficiently as a base-layer control for many decades. But with increased global competitiveness, process industries have been forced to reduce production costs in order to maximize profit. They must continuously operate in the most efficient and economical matter possible.

Most modern chemical processes are multivariable (i.e., multiple inputs influence same output) and exhibit strong interaction among the variables. Let us consider an operation of a boiler whose main function is to produce and deliver steam to downstream units or steam header at a specified temperature and pressure. The boiler has a drum with inlet water flow and is heated by fuel gas to produce steam. Now consider a situation where demand of the steam in downstream units increases, and it starts drawing more steam from the header. As a result, water level in the boiler will drop, vapor space above the water will expand, and consequently pressure and temperature will drop. Note that the water level in boilers is not independent and can affect the steam pressure and temperature. As a corrective action, if inlet water flow increases to control level, this will drop the boiler temperature. It will call for more heating and more evaporation, which will again lead to level drop. This demonstrates that there are very strong multivariable interactions among steam pressure, temperature, boiler level, and inlet water flow. Everything affects everything.

Now consider a conventional basic regulatory control scheme in a boiler where multiple single-input, single-output PID controllers are used for controlling the plant (multiloop control). Say, boiler level is controlled by inlet water flow, temperature is controlled by fuel gas flow, and boiler pressure is controlled by outlet steam flow. One basic shortcoming of PID loop controls is that they act as a single-input, single-output (SISO) controller in an island mode. For example, level controller will see and maintain only level with no idea what is happening with pressure and temperature. The same is true for the temperature controller, which will adjust the fuel gas base on temperature feedback and will not care for level. Now consider the previous situation, where the level starts dropping due to more drawing of steam from header. Level controller will increase inlet water flow, which will reduce the temperature. Temperature controller will increase fuel gas flow, which will again lead to level drop. Again, the level controller allows more inlet water flow to maintain level. There is a lack of coordination among the controllers, and they all act as unconnected islands. Neighboring PID loops can cooperate with each other or end up opposing or disturbing each other. This is due to loop interactions and is a serious limitation of PID regulatory controller. It is very important to understand the multivariable interactions in the chemical process plant and then try to develop model predictive control.

MPC usually stands for *model predictive control*. Model predictive control is used in multivariable processes where multivariable interactions among the process parameters are significant. However, MPC also stands for *multivariable predictive control. MPC* is used for both model predictive control and multivariable predictive control throughout this book. In industry, sometimes it is also called advanced process control or APC. Reader should appreciate that MPC stands for all of them in this book and fundamentally they refer to same model predictive control in a multi variable process environment.

Unlike the PID controller, MPC is a multi-input, multi-output (MIMO) controller. MPC receives all the inputs (e.g., temperature, pressure, level, fuel gas flow, inlet eater flow) and uses a predictive model to predict the output. As the name suggests, the heart of MPC is the predictive model. It then calculate the fuel gas flow, inlet water flow and other factors so that the controlled variables (temperature, pressure, level) are maintained at their set point or within their specified limit. The internal predictive model will account for all the multivariable interactions among the process parameters and adjust the manipulated variable accordingly. This is where MPC is more advantageous than multiloop PID controllers.

Let us consider another example of driving a car on the busy road. The car has two control outputs—namely, speed and direction. The manipulated variables are accelerator, brake, and steering. Now it must adhere to some safety limits—it cannot leave the road, it has to follow left or right side of the road depending on the country, it must maintain safe distance with other vehicles, and so on. There are some input constraints, also. For example, fuel injection cannot be increased rapidly to very large value.

Now consider two simple regulatory PID-type controls on the car. Let's say that direction is controlled by steering and speed is controlled by the accelerator, and these two controllers act independently without knowing each other. As it is well known, while turning on a curved road, speed will decrease—there are interactions between direction and speed. The two independent controllers are just like two drivers running the car, one only looks at direction and other only maintains the speed irrespective of direction, and there is no coordination between them. It would be a disaster, and cars cannot be drive in this manner. Here we need multivariable controller, which will simultaneously measure direction and speed and changes steering, accelerator, and brake simultaneously. This example is given to demonstrate the limitations of PID controllers in multivariable interactive process plant and MPC will be helpful over conventional multiloop PID controller.

1.3 What Is Multivariable Model Predictive Control?

In a process plant, it is only seldom that one encounters a situation where there is a one-to-one correspondence among manipulated and controlled variables. Given the relations between various interacting variables, constraints, and economic objectives, a multivariable controller is able to choose from several comfortable combinations of variables to manipulate and drive a process to its optimum limit and at the same time achieve the stated economic objectives. By balancing the actions of several actuators that each affect several process variables, a multivariable controller tries to maximize the performance of the process at the lowest possible cost. In a distillation column, for example, several tightly coupled temperatures, pressures, and flow rates must all be coordinated to maximize the quality of the distilled product.

Model predictive control is, as the name implies, a method of predicting the behavior of a process based on its past behavior and on dynamic models of the process. Based on the predicted behavior, an optimal sequence of actions is calculated. The first step in this sequence is applied to the process. For every execution period, a new scenario is predicted and corresponding actions are calculated based on updated information.

The real task of multivariable model predictive control (MPC) is to ensure that the operational and economic objectives of the plant are adhered to at all times. This is possible because the computer is infinitely patient, continuously observing the plant, and prepared to make many, tiny steps to meet the goals.

A definition that seems to provide the best overall description of what is intended is thus:

> MPC is the continuous and real-time implementation of technological and operational know-how through the use of sufficient computing power in a dynamic plant environment in order to maximize profitability.

1.4 Why Is a Multivariable Model Predictive Optimizing Controller Necessary?

Due to interactive nature of process variables, where change in one affects more than one related variable, it would be ideal to have a controller that is able to combine the operation of a number of single loop controllers. At the same time, this controller should also be able to choose, intelligently, a comfortable selective group of those variables whose manipulation will drive the object variable(s) to its or their optimum targets.

Interaction among process variables is a very common situation encountered in many process plants. Often, the selection of variables to be driven to their limits and extent of their manipulation is left to the subjective judgment, consequent of experience, of operator-in-charge. The selection is essentially a trade-off between variables to be driven to their limits. This is largely because of the complexity of interactions among variables. This judgment, while not wrong from a process or operational viewpoint, may not be in sync with company's objectives or market demands. Every operator recognizes the interaction among process variables. However, it is the impracticality of negotiating these variables to maintain an optimum condition at all times that forces operators to maintain the variables at a "comfortable" location, away from their constraints. A direct result of this is that the operation is never at its optimum point. This is where a multivariable controller steps in to perform.

Let us understand the concept of process interactions through a distillation column example. Figure 1.1 is a schematic of a distillation column (debutanizer column). In this scheme, there are two primary aims:

1) Reduce butane in column overhead.
2) Prevent slip of propane into butane at bottom.

The two concentration levels at top and bottom are themselves interactive. Further, the two are affected by reflux flow, reboiler steam, and variations in feed flow and quality. A variation in feed quality will change bottom composition sooner than it will affect top

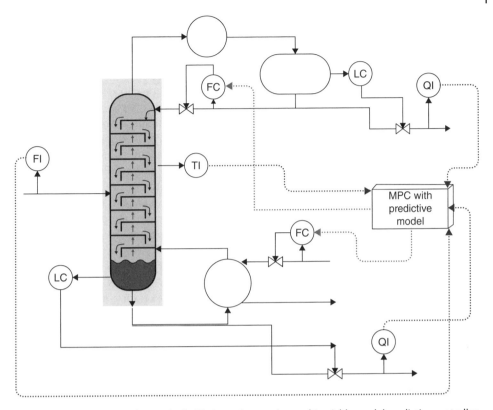

Figure 1.1 Flow scheme of a simple distillation column using multivariable model predictive controller

product. Once the controller senses the change in bottom quality, it will immediately calculate the changes to be made to reflux and reboiler steam flow rates to drive bottom quality to the set target. While arriving at the step values for reflux and steam flow rates that will drive bottom quality, the controller also takes cognizance of the effect of feed disturbance, reflux, and steam flow on top quality. This way, top quality is kept under control.

If the controls were single-loop controllers, one each for top quality and bottom quality, the column will swing because of interacting effects of change in steam flow and reflux flow on bottom and top qualities. A multivariable control system can also take into account the cost of applying each control effort and the potential cost of not applying the correct control effort. Costs can include not only financial considerations, such as energy spent versus energy saved, but safety and health factors as well.

Once the basic aim of maintaining quality is achieved, the focus can be shifted to achieving economic targets. It is possible to set a constraint (of the more expensive product in cheaper product), thereby minimizing loss of the more expensive product. While forcing the controller toward realizing this goal, it can also be asked to look into the possibility of reducing reflux flow and reboiler duties. All three goals set for the controller are dependent on one another. Yet, because it is a multivariable controller that has already been taught the effect of changing one variable on related variables, it keeps a check of the variables at each move.

1.5 Relevance of Multivariable Predictive Control (MPC) in Chemical Process Industry in Today's Business Environment

Business environment has drastically changed in last 20 years. Globalization, reduced profit margin, and cut-throat competition among process industries has changed the rule of the game. Making money by safely producing chemicals, petrochemicals, and oil refinery products are not sufficient to survive in today's business environment. Maximization of profit, continuous improvement of operation, enhanced reliability, and maximization of on-stream factor are buzzwords in today's CPI. Industries are shifting focus to energy efficiency, environment friendliness, and sustainability. Capacity maximization by exploiting the margin available in process equipment is no longer a luxury but a necessity. Maximize profit margin by reducing waste product, by increasing mass transfer and energy efficiency of equipment, and by pushing the process at their physical limit are the current trends of CPI. Reduction in oil price, decline in Chinese economy, growing instability in Middle East, and competition from US shale gas, for example, add new dimensions in business environment in recent years. Process industries are experiencing threats to their survival like never before. Stringent pollution control laws, enforcement of energy efficient process, involvement by government agencies, and decline of sales price of end products are some issues that force the process industries to look for new technological innovations to do business.

Traditionally, process controls were meant to control the process parameters within their safe boundary limit. This ensures a stable and safe operation. The introduction of sophisticated DCS system in CPI reduces manpower and production costs and increases the ability to monitor and operate the whole process from the control room. Traditional regulatory control (PID control) makes it possible to run the process at its set point given by DCS operator. However, optimization of process parameters to maximize profit depends on the individual operator understanding and efficiency. Online optimization of process, pushing the process at multiple constraints and deriving maximum profit margin, is no longer possible by traditional regulatory control. There is a need for higher-level control to oversee the scope of optimization. Multivariable process control fits that demand. The real task of MPC is to ensure that the operational and economic objectives of the plant are adhered to at all times.

1.6 Position of MPC in Control Hierarchy

Advance process control nowadays consists of several layers in a hierarchical fashion, as shown in Figure 1.2.

1.6.1 Regulatory PID Control Layer

The bottommost level, which is directly connected with the instruments (transmitters, valves) of the plant, is known as basic regulatory PID control level. PID control has single sensing element and single final control element. That's why they are called

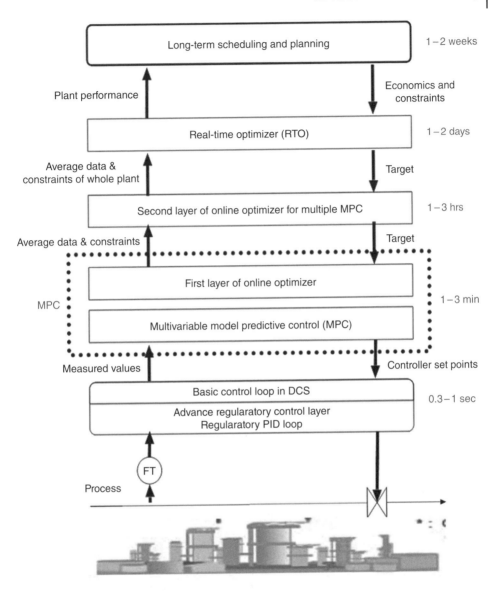

Figure 1.2 Hierarchy of plant-wide control framework

single-input, single-output system (SISO). Simple conventional temperature, pressure, flow control loops along with cascade control, and ratio control are fall under this level. A typical setup consists of a basic instrumentation layer in a DCS system that contains the PID controllers, which send outputs to the control valves in the plant. The primary goals of this base-level control are fast sensing the disturbance, fast disturbance rejection, and operational stability. Execution time is typically one second or less. PID controllers have been available for decades across chemical process industries and are considered the backbone of the industrial control system.

1.6.2 Advance Regulatory Control (ARC) Layer

The second layer just above the PID control layer is called the advanced regulatory control. The main feature of this control layer is that most of the time they are multi-input, single-output (MISO) systems. A few examples are pressure-compensated temperature, pressure, temperature, and density-compensated flow or mass flow, simple feedforward control based on auxiliary measurements, and override or adaptive gain control. Simple inferential process calculation-based control systems such as heater duty control and distillation tower flooding control also fall under this category. ARCs are typically implemented using function blocks or custom programming capabilities at the DCS level. Execution time is typically one to two seconds. Primary goals are enhanced stability and performance, simple inferential control, and feedforward actions along with feedback control.

1.6.3 Multivariable Model-Based Control

This type of control is characterized by simultaneous considerations of many controlled variables (CVs) and simultaneously adjusting many manipulated variables (MVs) based on model predictions. That's why this type of control is called a multi-input, multi-output (MIMO) system. The heart of the control is a dynamic model of the process that predicts future paths of control variables. Simultaneous control of more than one CV by coordinated adjustment of more than one MV is a main feature of this control layer.

An MPC is a multivariable model predictive controller that sits above a regulatory control implemented in DCS. In its turn, it receives its set point from a real-time optimizer. The tree can be extended upward to market forces. In a hierarchical sense, an MPC is a low-level controller. It takes cognizance of ground-level realities or constraints and directs basic controllers, already provided, toward an optimum target. The direction or targets can be set by operator or higher-level optimizers.

The principal aims of an MPC are as follows:

- Drive variables in a process to their optimum targets keeping in mind the interaction among variables.
- Effectively deal with constraints.
- Respond quickly to changes in optimum operating conditions.
- Achieve economic objectives.

Typical execution time is one to five minutes. MPC resides in a supervisory computer, which at regular intervals (typically once per minute) scans measured data from the DCS system and then writes set point to the regulatory control loop.

1.6.4 Economic Optimization Layer

Optimization layers sit at the top of multivariable control layer. Optimization again has three layers.

1.6.4.1 First Layer of Optimization

MPC provides first-level optimization. User normally specifies an economic objective function for the controller and assigns costs or values to its variables. Most of the time, most CVs are not required to control at specific set points; rather, they are required to be maintained between high and low ranges. This gives rise to degrees of freedom, and the optimizer translates into a direction to move the MVs. Steady-state optimizer calculates the objective function while obeying the constraints of CV and MV limits. Economic optimizer finds the best position (corner) in the solution space by using the linear program (LP) or quadratic program (QP) method of optimization. These constrained economic optimizers are typically used in conjunction with a multivariable control engine to implement the solution. The multivariable control layer and this first layer of optimizer together are known as model-based predictive control (MPC). Typical execution time is one to five minutes.

1.6.4.2 Second Layer of Optimization

This layer sits on the top of first layer of optimization and overlooks any potential opportunity that exists among the different units of a process plant. This optimization algorithm coordinates between different units of the plant, makes use of feedforward or disturbance variables, and tries to optimize the overall plant objective functions instead of individual unit objective functions with an aim to maximize the economic benefit of whole plant.

Each MPC application normally operates independent of the others. In order to optimize the operation of an entire unit, an online optimization system can be installed. MPC can accommodate *optimization by slogan* on each subplant over which it has control. These systems include calculations to the level of detail of, for example, tray-to-tray calculations in distillation columns. In the online optimization system, a profit function taking into account feedstock and energy cost and product prices is maximized for the unit. When calculating the optimum, all product flows from the unit are calculated, taking into account all known unit constraints and the heat integration in the unit.

Based on solutions of overall objective function, it provides set value to the MPC optimizer (first layer), which subsequently implemented it in the plant. Sometimes, running a particular unit suboptimally gives scope to increase profit in other units and this maximizes the overall profit. This particular feature is exploited in the multiunit optimizer. Typical execution frequency is one hour.

1.6.4.3 Third Layer of Optimization

Real-time optimizer (RTO) is the highest level of optimizer whose job is to optimally shift in operation priorities as per market demand and raw material availability. Just as, for example, fluidized catalytic cracking (FCC) unit of oil refinery is operated to produce different products like gasoline, ATF, and LPG. FCC unit can be operated under many following priorities, as per market demand:

- Maximization of gasoline or LPG product
- Maximization of ATF
- Maximization of profit
- Minimization of energy consumption

Now these market-driven operating goals must be achieved while operating conditions are also changing. Some examples of operating conditions changes are as follows:

- Changes in crude oil quality
- Changes in operating conditions due to catalyst performance degradation over time (low selectivity), heat exchanger fouling, or changes in separation efficiency in distillation columns

Now in a situation where objective function (operational priorities) and constraints (operating conditions and inputs) are changing dynamically, what is the best way to run the plant (FCC, for example)? Real-time optimizers (RTOs) address this issue. An RTO has complex algebraic and partial differential equations models of the whole system along with data reconciliation and it solves it dynamically (say, every 16 to 24 hours), while honoring all the operational constraints.

What should be the best operating points amid constraints and as per current market demand? RTO solutions answer this question and implement them in the plant. Solutions of RTO downloaded as external targets in the lower layer and all the lower-layer optimizers are subsequently implemented in the plant. The optimizer first calculates a base-case situation that identifies where the plant is currently operating. In a second stage, the optimizer will maximize a profit function, which results in setting the target values for control. The new target set points are downloaded every two to four hours. Via this technique, circumstances of changing economics, varying feed types, and modes can be handled such that a high number of multiple constraints can be met simultaneously and "hidden" opportunities from distant interactions in the plant can be exploited.

As already seen, advance process control today is a multilevel hierarchical control, solves a variety of problems, and requires large, multilevel technical skills starting with control engineering, planning and scheduling engineers, process engineering, and optimization experts. Today's process control systems not only keep the process parameters within their desired range but also dynamically move the plant in the most economically optimum zone. They keep recalculating and moving the plant while the economical optimum zone keeps moving from one place to another based on market demand and input constraints.

Figure 1.2 shows the full hierarchy of optimization/process control, as discussed in this section.

1.7 Advantage of Implementing MPC

Each type of controller realizes benefits for the user. As we move into realms of greater sophistication, the benefits are more. Figure 1.3 shows some typical benefits associated at each level of control.

Advanced control and optimization benefits are field proven and generate high rates of return on investment. Figure 1.4 shows some typical areas where MPC is used to bring more profit:

- *Increased throughput:* A typical 2 to 4 percent throughput increase is very common in industry after MPC implementation. This throughput increase is done while obeying different constraints and limits.
- *Better management of constraints:* As MPC has internal dynamic predictive model, it predicts the CVs future value and try to maintain them very near to constraints. In industry, the following constraints are reported to generate high returns:
  - Furnace capacity
  - Column flooding and hydraulic limits
  - Overhead condenser capacity limitation
  - Less flaring

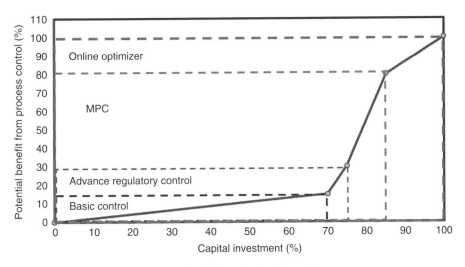

Figure 1.3 Expected cost vs. benefits for different levels of controls

- *Improved yields:* MPCs stabilize and optimize the process parameters by their predictive capability, which ultimately helps to improve yields, catalyst selectivity, etc.
- *Reduced energy usage:* MPCs constantly try to reduce the steam and fuel gas —i.e., energy input to the process by its optimization capability.
- *Decreased operating costs:* MPCs always move the process toward its most economic operating zone and hold the process at multiple constraints. It makes it possible to lower pressure operation, adjust fuel gas to oxygen ratio, and make more severe flushing, produce less slops, and so on, and thus continuously decrease operating cost of the unit.
- *Lower pressure operation:* Generally, a distillation column operated at lower pressure means better separation and lower energy costs. Column operating pressure is dictated by cooling water temperature available to cool light components coming from top. APC can be configured to keep a constant watch on column top pressure or reflux temperature, allowing the column to be operated at minimum possible pressure.
- *Less flue gas oxygen:* A typical case of multivariable controller. Target heater outlet temperature is maintained by simultaneously adjusting the fuel rate and airflow rate (or excess oxygen). More oxygen or air implies lower hearth temperatures or higher fuel consumption for the same heater outlet temperatures. Very low excess oxygen implies inefficient combustion or fuel loss.
- *Reduced slop make:* APC not only minimizes producing better-quality product but also avoids making off-spec product. Hereby, it reduces slop make.
- *Improved quality consistency:* MPCs make less quality giveaway against specification. When quality is a critical parameter, it is the general tendency to operate the plant at a condition where the average quality is better than necessary. It is cost to pay for being safe. With APC in place, the swings are reduced. This automatically translates into money due to reduced give away.
- *Increased production of more valuable products:* In refineries where high-value products are occasionally blended with low-value products, it is imperative that only

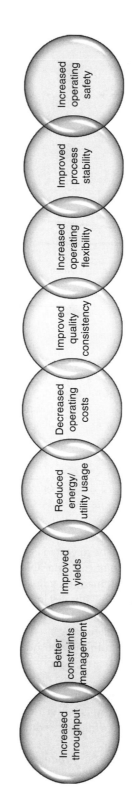

Figure 1.4 Typical benefit of MPC

the requisite quantity of high-value product be used. Any excess use is loss. APC, with its smoothening effect, minimizes these losses while maintaining quality target.

- *More stable product quality:* MPCs are able to maintain more stable product quality for blending/downstream unit.
- *Increased operating flexibility:* MPC uses coordinated movement of multiple MVs to control multiple CVs. It uses the impact of disturbance variables on CVs as a feedforward variable. All these are used to give more flexibility to operate the process plant.
- *Easier and smoother operation with less operator intervention:* This will reduce their stress while more rapidly returning the process to normal operations, and it provides a smoother, faster, and better transition between different feed rates.
- *Reduced time off-spec during grade change of polymer or crude changeover.*
- *Improved process stability:* MPC uses its inherent model to stabilize the process first. Typically it reduces the standard deviation of key process parameters to 50 percent of pre-MPC value.
- *Reduced utility consumption (stripping steam, driving steam, air blower, etc.):* When a column operates at lower pressures, reboil requirement is reduced.
- *Increased operating safety:* This is in direct relation to uptime of MPC. Higher uptimes signify less operator intervention in process. This reduces chances of error making process inherently safer.
- *Implementation of APC/MPC:* In terms of blend property control and blend ratio control in the blending area, implementation usually results in the following:

  - Reduced quality giveaway
  - Optimum use of the more valuable components and additives
  - Reduced reblending requirements
  - Better usage of the available tankage
  - Typical benefits range from US¢ 10–20/bbl. of product
  - Reduction of reformer processing severity due to the reduced requirement for octane achieved by better blend property controls
  - Increased blender throughput as the need for reblends is reduced or eliminated
  - Potentially huge savings may be realized through reduction in operating incidents with the use of blending APC solutions

1.8 How Does MPC Extract Benefit?

Process conditions that can be improved dramatically through MPC application are listed in this section.

1.8.1 MPC Inherent Stabilization Effect

MPC first stabilizes the process. If a process is unstable or has poor control on its key parameters, there is always a scope to potentially lose some of the economic benefits that could have been captured if a well-controlled process were available. MPC uses feedforward information, best use surge volumes available in the plant, and continuously taking small incremental steps for preventive and corrective actions to stabilize the process.

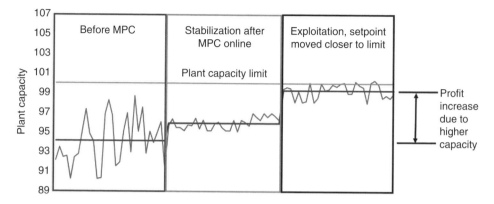

Figure 1.5 MPC stabilization effect can increase plant capacity closer to its maximum limit

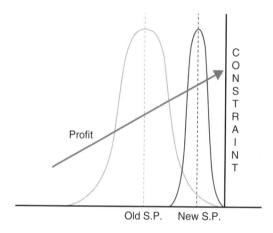

Figure 1.6 Reduced variability allows operation closer to constraints by shifting set point

The immediate effect of making a move while keeping in view the amplitude of move and its consequence is a smoother operation. MPC takes advantage of this. It pushes its controlled variables toward their constraints. This phenomenon is explained in Figure 1.5.

The real benefit from MPC is obtained not just by reducing variations but also by operating closer to constraints, as illustrated in Figure 1.6.

1.8.2 Process Interactions

Modern chemical plants are highly integrated and have large interactions among process variable. Nowadays, due to intense global competition, plants were designed with large recycle, heat integrated, less design margin, and more. This leads to large interactions and makes different process parameters highly dependent on each other. Normal PID regulatory controllers cannot see these interactions and act as independent controller in island mode. Due to this highly interactive environment, tighter control is impossible in PID controller level. As MPC knows about the interaction through its MV-CV-DV models, it is capable to compensate the interactions with its prediction capability. Thus, MPC allows optimal and tighter control possible.

1.8.3 Multiple Constraints

Any chemical process has multiple constraints or limits. These limits are posed by equipment hardware limits (like flooding, furnace tube maximum temperature limits, etc.) and process safety limits (like maximum allowable oxygen concentration to avoid flammable region, etc.). This is shown in Figure 1.7. For an efficient operator, it is impossible to run the process at its limit all the time. MPCs predict onset of such constraints and push the process to hit multiple constraints. MPCs maximize profit function reflecting most favorable combination of conflicting constraints. MPC pushes the process out of the comfort zone of individual operator toward the optimum performance while still honoring the plant safety and operating constraints like high and low limits (set points) for CVs, product composition specifications, metallurgical limits, valve output. All constraints are considered and accounted for in the overall control and optimization strategy.

Unmeasurable targets: MPCs can control calculated variables (commonly called soft sensor or quality estimator)—that is, it can control target or constraints that are not directly measured. Examples are distillation tower flooding, tray or down comer loading, compressor efficiency, impurities in product, and so on.

Large delay process: Many commercial chemical processes have large delay in their responses. These delays may be due to large process inventory (like temperature rise after increasing reboiler steam), process itself (e.g., catalyst selectivity change after changing inhibitor or activator feed rate) or may be due to analyzer dead time. As these delays are captured during step test and incorporated inside the MPC model, MPC is better positioned to anticipate the delayed response and control accordingly.

Difference in product value in a multiproduct plant: for a multiproduct plant, product yields can be changed dynamically as per market conditions by utilizing MPC economic

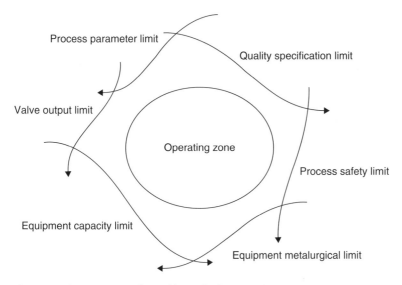

Figure 1.7 Operating zone limited by multiple constraints

objective value maximization. MPCs can exploit its internal optimization capability if such condition exists.

Different mode of operations: MPCs can be modeled to accommodate different regime or mode of operation such as different polymer grade or different quality crude oil processing. MPCs can be used to quickly switch one mode to other online without much disturbance in the process.

Plant load maximization: When all the constraints are within limit, MPC tries to maximize the plant load continuously with small incremental steps. Due to its prediction capability, MPC will do that without violating any major constraints.

Optimization of load distribution between different equipment: As the MPC model has good understanding of constraint limit, it is always possible to allocate feed in such a way in different units so that the most efficient unit utilization can be maximized. It can also be possible to balance column tray load between rectification and stripping section.

Ambient conditions change: Ambient conditions like cooling water temperature and air temperature vary between day–night and summer–winter. For example, low cooling water temperature at winter can increase condensing load in a distillation column condenser, thus providing extra opportunity to increase load in column assuming that condenser heat duty pose a constraint in summer time. MPC is well equipped to exploit any such spare degree of freedom.

Figure 1.8 shows how opportunity is lost in a typical plant by operator actions. MPC can act on this lost opportunity and bring more profit as it is continuously optimizing the plant on a minute-by-minute basis.

- *Use of soft sensor:* Sometimes reliable analyzers may not exist for quality estimation of a major product. In such situations, soft sensors can be made that will calculate the quality parameters based on other available process parameters. MPC can use these soft sensors readings as a controlled variable and can control the quality based on real time predictions.

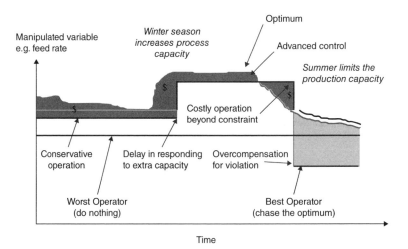

Figure 1.8 Opportunity loss due to operator action

- *Excessive operator intervention area:* During grade change of polymer, excessive operator interventions sometimes needed to carry out sequence of predetermined action. MPC can be utilized in such situations to automate complex series of action consistently without any error.

1.8.4 Intangible Benefits of MPC

One of the major intangible benefits of MPC is to change the mindsets of operators. Before MPC, operators used to monitor process parameters like flow, level, temperature, and pressure. After MPC, operator monitor performance parameters like yields, selectivity, specific energy, efficiency, and constraints. Thus, operation staff can better align themselves to company objective such as maximizing profit and pushing the process to its limit. The obvious benefits are that the process becomes more flexible, there is less switchover timing of crude/polymer grade, and there is less off spec production.

Intangible benefits are harder to quantify but can be quite significant, depending on the project under evaluation:

- *Improved process safety:* MPC works as a process watchdog and is capable of identifying any problem beforehand, as it has predictive capability.
- *Improved operator effectiveness:* After MPC implementations, operators can focus on key operating parameters and plant constraints rather than focusing on every temperature, pressure, and flow value.
- *Reduced downstream unit variability:* It has been reported in industry that fewer process upsets are seen in downstream units after MPC implementations.
- *Better process information:* As MPC works on plant economics and plant constraints, operators and engineers have better process understanding after MPC implementation. They start to look at the unit as a profit generator rather than as a simple process unit.

All of these benefits cannot be measured by monetary terms, but they have profound long-term intangible effect on plant economics.

1.9 Application of MPC in Oil Refinery, Petrochemical, Fertilizer, and Chemical Plants, and Related Benefits

MPC has been applied to almost all types of chemical plants since its inception in 1970. MPC application started in chemical plants and oil refineries and then spread to other chemical plants and slowly other branches of engineering. Currently, it is used in robotics application, space application, and biochemical plants. Normally, process responses in chemical plants are considered slow; that's why it started with chemical plants in 1970s, when computers took a long time to calculate online optimization solutions. Over the years, MPC technology passed through many structural modifications, and now it is considered as a mature technology with large applications in chemical, petrochemical, and refineries across the globe. Bowen (2008) describes an application of advance process control, RMPCT of Honeywell, in crude distillation unit. Similarly, McDonald (1987) gives an application of dynamic matrix control to distillation towers.

Wang (2002) describes application of multivariable predictive control technology in atmosphere and vacuum distillation unit. Other good references of various application of MPC in industry are Clarke (1988); Ordys (2001); and Cremaschi (2005).

The typical payback period of an MPC project is six months. Some actual implementation payback periods are given in Table 1.1. Industries across the globe have reported huge profits due to MPC implementations. Table 1.2 and Table 1.3 give some ballpark benefit numbers reported from chemical industry and refinery, respectively (Honeywell Inc. 2008).

To give the idea of penetration of MPC in the chemical industry, let us consider what one major MPC vendor alone has implemented across globe (refer to Figure 1.9). Figures inside the picture are dynamic and continuously changing. They are for illustration

Table 1.1 Typical Payback Period of MPC

Plants	Typical Payback Periods
Oil and gas	1–2 months
Refining	3–6 months
Chemicals	3–6 months
Petrochem	4–6 months
Pulping/paper	6–8 months
Industrial power	10–12 months

Table 1.2 Typical Benefits of MPC Implementation in CPI

Petrochemicals	Benefits (per year)
Ethylene	2–4% Increase in production
VCM	3–5% Increased capacity/1–4% yield improvement
Aromatics (50KBPD)	3.4M–5.3M US$
Chemicals	
Ammonia	2–4% Increased capacity/2–5% less energy/ton
Poly olefins	2–5% Increase in production/Up to 30% faster grade transition
Oil & Gas	
Upstream production	1–5% Increase in production
Industrial utilities	
Cogeneration/Power systems	2–5% Decrease in operating costs
Pulping	
Bleaching	10–20% Reduction in chemical usage
TMP (Thermos Mechanical Pulping)	$1M–$2M

Table 1.3 Typical Benefits of MPC implementation in Refinery

Refining	Benefits ($0.01/bbl.)	Benefits (US$/yr)
Crude distillation (150 KBPD)	5–13	2.7M–7.0M
Coking (40 KBPD)	15–33	2.2M–4.8M
Hydrocracking (70 KBPD)	13–30	3.3M–7.6M
Catalytic cracking (50 KBPD)	13–30	2.4M–5.4M
Reforming (50 KBPD)	10–26	1.8M–4.7M
Alkylation (30 KBPD)	10–26	1.1M–2.8M
Isomerization (30 KBPD)	3–17	0.3M–1.8M

Figure 1.9 Advance control implementations by one of the major MPC vendors

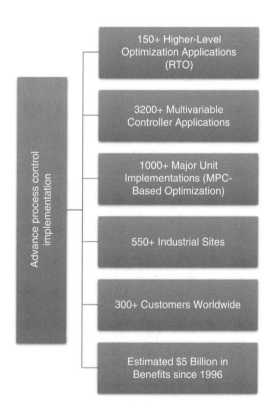

purposes only. Other major MPC vendors like AspenTech and Shell Global solutions also have a similar large number of implementations across globe.

MPC application has now spread across the whole spectrum of chemical process industries. Let us consider what Honeywell alone has achieved, as shown in Figure 1.10. Figures inside the picture are dynamic and continuously changing. They are for illustration purposes only.

Figure 1.10 Spread of MPC application across the whole spectrum of chemical process industries

References

Bowen, G. X. D. F. X. (2008). Application of Advanced Process Control, RMPCT of Honeywell, in DCU. *Process Automation Instrumentation*, 4, 013.

Clarke, D. W. (1988). Application of generalized predictive control to industrial processes. *IEEE Control Systems magazine*, 8(2), 49–55.

Cremaschi, R. A., & Perinetti, J. A. (2005). Implementing multivariable process control. *Petroleum Technology Quarterly*, 10(1), 115–119.

Honeywell Inc. (2008). Layered Optimization: A low-risk, scalable approach to driving sustainable plant wide benefits. Optimization Solution White Paper from Honeywell, Inc., USA, April 2008.

McDonald, K. A., & McAvoy, T. J. (1987). Application of dynamic matrix control to moderate-and high-purity distillation towers. *Industrial & Engineering Chemistry Research*, 26(5), 1011–1018.

Ordys, A. W. (2001). Predictive control for industrial applications. *Annual Reviews in Control*, 25, 13–24.

Wang, C. M., & Lei, R. X. (2002). Application of multivariable predictive control technology in atmosphere and vacuum distillation unit. *Petrochemical Technology and Application*, 20(5), 321–323.

2

Theoretical Base of MPC

2.1 Why MPC?

Profit maximization is the buzzword of today's chemical process industry. In today's competitive business world, the main purpose of any process unit is to maximize the profit on a USD/hour basis and sustain it. Some ways to maximize profit are to maximize plant throughput while obeying all design and safety limitations, minimize raw material and utility consumption, reduce production cost by maximizing process efficiency (like selectivity, yield, specific consumption), increase plant reliability, and obey all the constraints imposed by licensor, designer, equipment manufacturer, pollution control authorities, and others so that profit-making production process can be sustained for longer periods. In still other cases, there is a trade-off between increased throughput and decreased process efficiency such that process optimization is needed.

For any process unit, an acceptable operating region is bounded by various limits, which fall into one of the following categories:

- Safety limits (e.g., maximum allowable oxygen concentration in hydrocarbon mixture to avoid flammable mixture)
- Actuator limits (e.g., a valve is either open or closed)
- Equipment limits (e.g., the maximum vessel working pressure or temperature)
- Operational constraints (e.g., a compressor surge limit, a tower differential pressure to avoid flooding)
- Product quality constraints (e.g., upper limit on product impurities)

Normally, panel operators and production engineers try to operate the plant at the center of this acceptable operating region, far from any constraints. The reason is simple: The panel operator gets the maximum amount of time to respond to disturbances before it drives the process beyond the acceptable operating region. This gives the operator some flexibility in operation. However, to get maximum profit from the process, it has to push several constraints or limits, and usually this most economic operating point lies at the edge of the boundary limit (see Figure 2.1). This is true whether the unit is run for maximum throughput or not.

Multivariable Predictive Control: Applications in Industry, First Edition. Sandip Kumar Lahiri.
© 2017 John Wiley & Sons Ltd. Published 2017 by John Wiley & Sons Ltd.

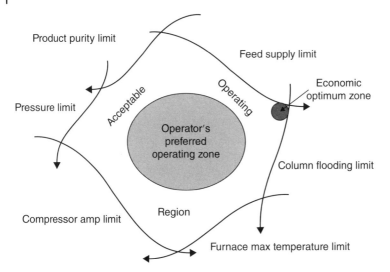

Figure 2.1 Optimum operating point vs. operator comfort zone

However, operating the process at multiple constraints is not an easy task and has several shortcomings:

- *It takes away the operation flexibility from the operator.* Corrective actions must be taken very quickly in case of disturbances such as feed quality change or cooling water temperature change; otherwise, the process will go to unacceptable operating regions, which may have negative impact on product quality or throughput.
- *Some of the manipulated variables will be at their limit in this economic operating zone.* The operator will thus not have the luxury to use them to compensate for disturbances.
- *Due to the highly interactive nature of the process, changes in one manipulated variable (MV) will affect several controlled variables (CVs).* It is difficult for the human operator to track these impacts. The process will thus be prone to enter into unacceptable operating regions.
- *The location of optimum operating points is not constant.* They change their position from one edge to the other during the course of day due to factors such as feed change, utility change, and ambient conditions changes.

Panel operators equipped with a traditional PID control loop cannot track the movement of the economic operating point minute by minute and are also not expected to coordinate different MV moves to drive and maintain the process at several constraints limit. That's why there is always a margin between the current and optimum economic point. So financial opportunities always exist to exploit this margin.

If a control scheme could be created to detect this economically optimum operating point at each controller execution (typically every minute), drive the process operation to this point, and operate the process in a stable manner at this point in the face of disturbances, the economic potential could be realized. This is the issue that the MPC software has been created to solve.

2.2 Variables Used in MPC

There are three types of variables in MPC, which are frequently used throughout this book and are explained below.

2.2.1 Manipulated Variables (MVs)

Manipulated variables are the control handles on the process. These are the independent variables whose conditions are manipulated (changed) to control the CVs. MPC adjusts MV values to achieve control and optimization objectives. The controller never moves an MV outside its limit.

Examples of MVs
- Reflux flow
- Reboiler steam flow or duty
- Overhead pressure
- Feed temperature
- Feed flow
- Compressor speed
- Heater fuel gas pressure

2.2.2 Controlled Variables (CVs)

Controlled variables are the process conditions to be controlled (that is why they are also called PVs, process variables). The bottom temperature of a distillation column can be CV.

Examples of CVs
- Temperature
- Pressure
- Delta pressure
- Inferential calculations
- Concentration measured by analyzers
- Valve position

2.2.3 Disturbance Variables (DVs)

Disturbance variables are measured disturbances (changes) in the process that influence the CVs, but that are not under MPC control. DVs often cannot be moved independently. Cooling water temperature, ambient temperature can be DV.

Examples of DVs
- Feed flow
- Feed temperature
- Feed composition
- Ambient temperature
- Cooling water temperature

2.3   Features of MPC

The MPC software package is based on technology designed for controlling highly interactive, multivariable processes at several constraints simultaneously.

This technology operates at the supervisory control level. Typical execution intervals are on the order of one to four minutes, and the MPC controller normally manipulates the set points of regulatory controllers.

Following are characteristics of the MPC controller, which demonstrate the power of the multivariable controller.

2.3.1   MPC Is a Multivariable Controller

As most of the chemical processes are highly interactive and multivariable, changes in one independent variable will affect several dependent variables. Instead of trying to correct the error of one dependent variable from its set point, MPC algorithms consider all the dependent and independent variable interactions and then formulate the move plan to reduce overall CV error. The MPC controller "knows" the system is multivariable and use its internal calculation to deal it.

2.3.2   MPC Is a Model Predictive Controller

At the heart of the MPC algorithm is the dynamic process model, which contains all significant interactions between independent and dependent variables. Normally, these models are generated from data during a step test. The model is then used to predict the open-loop behavior of the controlled variables for a period of time into the future, which is long enough to allow the effects of all past changes in the manipulated and disturbance variables to settle out. This settling time is called the steady-state time of the process.

This allows MPC to anticipate, or predict, future constraint violations, so that control action can be taken well in advance of the actual violation. This future prediction is reconciled with actual controlled variable measurements at each control cycle in order to eliminate model mismatch. This model predictive capability allows for the modeling of processes with unusual dynamics, such as long dead times or inverse responses.

2.3.3   MPC Is a Constrained Controller

Normally, panel operators set the upper and lower limits of all CVs and MVs based on the process requirements. MPC always honors the limits on MVs. Simultaneously MPC always tries to honor the limits of CVs, too, but in case of difficulty to honor all of them it has definite plan to sacrifice temporarily some of the least important CVs. Constraints on both controlled and manipulated variables are dealt with explicitly. When MPC plans how a disturbance is to be compensated for, it calculates current and future moves in the manipulated variables. When calculating this move plan, MPC algorithms take care so that the moves do not violate the upper and lower limits.

Otherwise, the controller might plan moves that could not be implemented. MPC explicitly handles these future constraints, ensuring that the calculated plan is one that can be implemented.

2.3.4 MPC Is an Optimizing Controller

Most commercially available MPC software MPC algorithm incorporates a linear program (LP) or quadratic program (QP) to solve a linear steady-state optimization problem for the most economic operating point at every execution of the controller. This optimization problem uses the predicted steady-state values of the controlled variables and the current values of the manipulated variables, along with cost information on raw materials, products, and utilities. These values are used to calculate the optimum steady-state operating point that satisfies the limits on all manipulated and controlled variables.

This steady-state operating point is imposed on the control problem, in which a dynamic optimization problem is solved. This dynamic optimization problem minimizes controlled variable error away from the LP/QP calculated steady-state operating point, while preventing manipulated variable limit violations.

2.3.5 MPC Is a Rigorous Controller

MPC assumes that the processes can be represented by a set of linear differential equations or algebraic equations (typically true for DMC plus software). In MPC, no assumptions are made about the form of the model; any form is allowed. MV-CV relationships can be obtained directly from the data. Some MPC technologies make assumptions about the form of the model, limiting it to first- or second-order dead time or an equivalent Laplace transform model (typically true for RMPCT software). This permits the most accurate prediction of future controlled variable values.

MPC provides the steady-state linear program (LP) or quadratic program (QP) to solve for economic optimum operating point, based on the costs and constraint values provided. The dynamic control problem minimizes the current and future error in each control variable all the way to steady state. MPC also provides constraint handling for future values of the manipulated variables.

2.4 Brief Introduction to Model Predictive Control Techniques

MPC controllers are designed to drive the process from one constrained steady state to another. They may receive optimal steady-state targets from an overlying optimizer, as shown in Figure 1.2 or they may compute an economically optimal operating point using an internal steady-state optimizer. There are five general objectives of an MPC controller, in order of importance:

1) Prevent violation of MV and CV constraints.
2) Drive the CVs to their steady-state optimal values (dynamic output optimization).
3) Drive the MVs to their steady-state optimal values using remaining degrees of freedom (dynamic input optimization).
4) Prevent excessive movement of MVs.
5) When signals and actuators fail, control as much of the plant as possible.

The translation of these objectives into a mathematical problem statement involves a number of approximations and trade-offs that define the basic character of the controller. Like any design problem, there are many possible solutions; it is no surprise that there are a number of different MPC control formulations. Different vendors use different dynamic strategy to implement the control philosophy. Various researchers and MPC experts come up with different formulations of MPC strategy and underlying mathematics. Those details can be found in Camacho 2013; Qin 2003; Morari 1999; Garcia 1989; Cutler 1983; Yousfi 1991; Badgwell 2015; and Kouvaritakis 2016.

It is important to understand the strategy first, then the underlying mathematics, and at last the difficulty in implementing the strategy. The underlying mathematics are not discussed in detail here but are available in any advanced process control reference book (Camacho 2013; Badgwell 2015; Kouvaritakis 2016).

The following sections explain the underlying concept.

2.4.1 Simplified Dynamic Control Strategy of MPC

Dynamic control strategy can be explained by Figure 2.2. The MPC scheme started with plant DCS, which is connected with different flow, temperature, level, and pressure transmitter in actual plant. Prediction modules take CV, MV, and DV data from plant DCS as input and predict the steady state value of CVs, as shown in Figure 2.2. This prediction module utilizes MPC internal predictive model for this purpose. A steady-state optimization module utilizes this CV prediction as input to the module. Other required inputs are MV, CV limits (given by operator), tuning constants (given by control engineer), and MV and DV values from DCS, as shown in Figure 2.2. It then calculates steady state optimum targets for CV and MV. The next step is to formulate a dynamic strategy so that these targets can be achieved. To do that, these targets are fed to dynamic control module to calculate MV movement value so that the controller

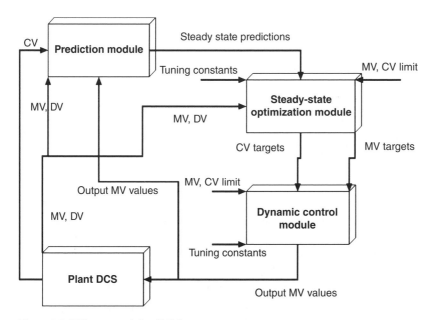

Figure 2.2 Different module of MPC

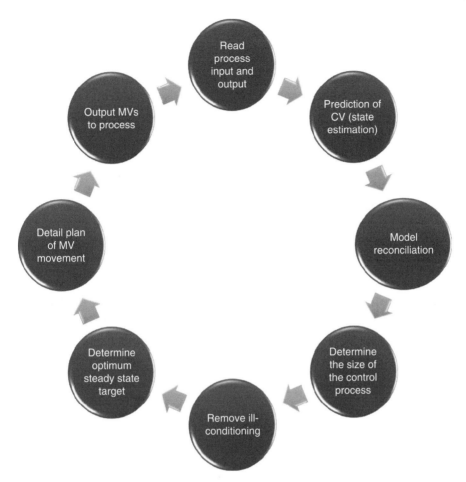

Figure 2.3 A general MPC calculation

can bring the plant to their respective optimum steady-state target. Dynamic control modules require MV, CV limits, and tuning constants as inputs to formulate this strategy. Dynamic control modules then calculate a series of MV movements, which then write back to PID controller set points (as MV) for implementations in an actual plant. This simplified strategy is shown in Figure 2.2.

The next step is to understand the various jobs performed during MPC calculations with a little bit detail. Different steps in MPC calculations are shown in Figure 2.3, and explained next.

2.4.2 Step 1: Read Process Input and Output

The first step is to read the current values of process inputs (DVs and MVs) and process outputs (CVs). After reading the CV, MV, and DV values, their validity is checked to gauge whether those values are good to use for subsequent calculations.

2.4.3 Step 2: Prediction of CVs

This step is performed in prediction modules, as shown in Figure 2.2. This step essentially means to find out where the process is now, and where it is heading. MPC do this with the help of its process models. Before implementing dynamic control strategy of MPC, it is necessary to build a dynamic model of the process.

2.4.3.1 Building Dynamic Process Model

At the heart of the MPC calculations is the dynamic predictive model. These models are data-driven dynamic models generated from the data during step test. In step testing, each of the independent variable is perturbed and their dynamic effects on CVs are collected. MPC software inherent system identification modules analyze these data and build MV-CV relationships from them. These relationships form dynamic, multivariable model of the process and capture all significant interactions between variables.

Consider a simple distillation column with six CV, three MV, and one DV, as shown in Figure 2.4. Descriptions of CV, MV, and DVs are given in Table 2.1.

Figure 2.5 shows the schematic of the model for the simple distillation column in matrix form. Each box represents the response in time of the dependent variable (CV) to a step change at time zero of the corresponding independent variable (MV), while all other independent variables are held constant (open-loop step responses).

In simple terms, each curve represents the effect of a change in an independent variable on that dependent variable.

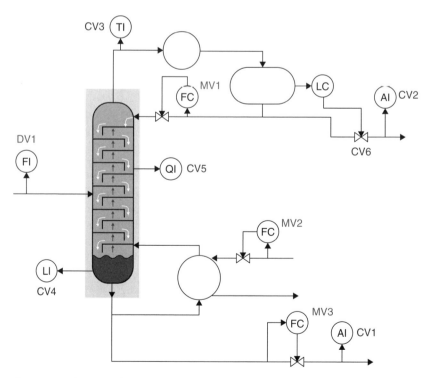

Figure 2.4 Schematic of distillation column

Table 2.1 Description of CV, MV, and DV in a Simple Distillation Column Shown in Figure 2.4

Variable	Type of Variable	Description
CV1	Control variable	Bottom stream purity measured by online analyzer, ppm
CV2		Top stream purity measured by online analyzer, ppm
CV3		Column top temperature, C
CV4		Column bottom level, %
CV5		Flooding% measured by soft sensor, %
CV6		Top control valve opening, %
MV1	Manipulated variable	Reflux flow, m3/hr
MV2		Steam flow, m3/hr
MV3		Column bottom flow, m3/hr
DV1	Disturbance variable	Feed flow, m3/hr

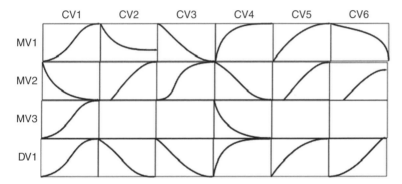

Figure 2.5 Model of distillation column

Once the dynamic model has been obtained, this model is used to form and maintain a *prediction of future behavior* of the controlled variables in the process. This is done in the prediction module of Figure 2.2.

MPC predict the future trajectory of the control variables over a finite time horizon using history of the past manipulated and disturbance variable changes. All independent variables changing up to one steady state time into the past are considered, since all of these changes still have an effect on the system. Changes that occurred more than one steady state time into the past need not be considered, since they no longer affect the system.

Since the model curves represent the effects of independent variable changes on dependent variables, these independent variable changes can be applied to the model to generate a future prediction for each dependent variable. These predictions extend from the current time out to one steady state time into the future. These dependent variable predictions are updated at each execution of the controller, and are reconciled with the actual dependent variable measured values to eliminate model mismatch.

2.4.3.2 How MPC Predicts the Future

To understand the concept, let us begin with a simplified case of SISO model prediction. Figure 2.6 shows the CV prediction for past two MV change (+1 unit at $t = 1$ and −2 unit at $t = 2$). The step response curve can be expressed as a matrix form given in equation 2.1 for a step change of one unit in MV.

Left-hand column vector (A) contains the model coefficients at successive scans intervals; these coefficients can be found from the step response curve. The 1×1 ΔMV matrix contains the input step change (ΔMV). The right-hand column vector (Y) defines the predicted response of the model, obtained by multiplying successive rows of the A matrix by the magnitude of ΔMV. Equation 2.2 shows a combination of two sequential changes in the MV. The predicted CV value at the right-hand side can be calculated by following the linearity and superimposition principle. In this way, by simple matrix operation future CV values can be predicted. The attraction of this approach is the ease with which this model representation can be cast into a matrix formulation.

$$A \times \Delta MV = Y$$

$$
\begin{bmatrix}
0 \\
0 \\
0.68 \\
0.99 \\
1.37 \\
1.58 \\
1.85 \\
1.87 \\
\ldots \\
2.00
\end{bmatrix}
\begin{bmatrix} 1 \end{bmatrix}_{t=1}
=
\begin{bmatrix}
0 \\
0 \\
0.68 \\
0.99 \\
1.37 \\
1.58 \\
1.85 \\
1.87 \\
\ldots \\
2.00
\end{bmatrix}
\tag{2.1}
$$

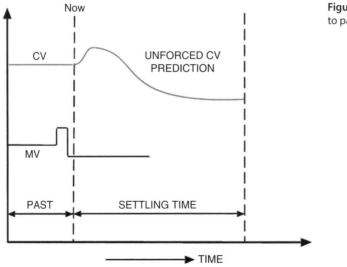

Figure 2.6 CV prediction due to past MV change

$$
\begin{bmatrix} 0 \\ 0 \\ 0.68 \\ 0.99 \\ 1.37 \\ 1.58 \\ 1.85 \\ 1.87 \\ .... \\ 2.00 \end{bmatrix} \begin{bmatrix} 1 \\ -2 \end{bmatrix} = \begin{bmatrix} 0 \\ 0 \\ 0.68 \\ 0.99 \\ 1.37 \\ 1.58 \\ 1.85 \\ 1.87 \\ .... \\ 2.00 \end{bmatrix} + \begin{bmatrix} 0 \\ 0 \\ -1.36 \\ -1.98 \\ -2.74 \\ -3.16 \\ -3.70 \\ -3.74 \\ .... \\ -4.00 \end{bmatrix} = \begin{bmatrix} 0 \\ 0 \\ -0.68 \\ -0.99 \\ -1.37 \\ -1.58 \\ -1.85 \\ -1.87 \\ .... \\ -2.00 \end{bmatrix}
$$

[2.2]

2.4.4   Step 3: Model Reconciliation

Since the step test models are data driven, they are not 100 percent accurate. Because the models are not perfect and by definition do not take account of unmodeled disturbances, there is a need to reconcile the model-based predictions to the process measurements and then feed this information back to update the model predictions into the future.

The predictor on the LHS of the picture predicts where the CVs are going. The model is really a subset of this, predicting where the CVs should be at the current scan. Figure 2.7 shows the essence of this approach. The error between the current and predicted value for the CV is used as a bias estimate to update the future trajectory of the CV. The error is filtered to eliminate noise in some MPC packages before being used as a bias update.

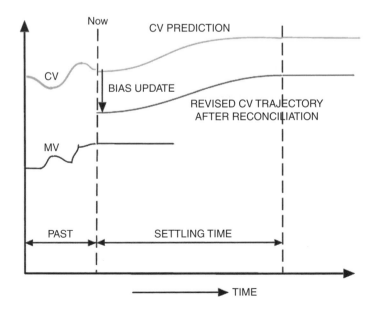

Figure 2.7 Model reconciliation and bias update

The model bias is assumed to stay constant over the time horizon—that is, it is a systematic error to the models rather than a random error.

For stable processes, all of the algorithms of majority of MPC vendors use the same form of feedback, based on comparing the current measured output to the predicted output:

$$b_k = y_k^m - y_k \qquad [2.3]$$

The bias term is then added to the model for subsequent predictions:

$$y_{k+j} = g(x_{k+j}) + b_k \qquad [2.4]$$

2.4.5 Step 4: Determine the Size of the Control Process

After the process state has been estimated, the controller must determine which MVs can be manipulated and which CVs should be controlled.

For a real process, it may happen that all the MVs and CVs (as per design) are not available for control at a particular point of time. There are four reasons for this:

1) Panel operator may deactivate one or a particular MV purposefully so that it is not available for control purposes.
2) Some of the MVs may be at their limit and not available for control.
3) Indication of some of the CVs may be faulty and operator put them out of MPC.
4) Some of the control valves may be at their 100 percent opening and saturated. Related MV may be not available for control.

All of these conditions are dynamic, and the panel operator can change the availability of MVs and CVs at any point of time. In general, the shape of the subprocess changes in real time. Hence, it is important to know which MVs and CVs are available at each control execution step. From these decisions, a controlled subprocess is defined at each control execution.

2.4.6 Step 5: Removal of Ill-Conditioned Problems

Control of interacting variables requires coordinated moves of a number of MVs. The controller determines how to make these moves by simultaneously considering the effects of all MVs on all CVs. The controller's knowledge of these effects is based on the process model.

If the model contains significant error, the effects of the control actions on the process differ from the effects predicted by the controller, and the quality of control degrades to some extent.

The problems caused by model error are more pronounced for processes that have large interactions between CVs that are assigned set points or are at constraints most of the time.

Unavoidably, all models contain some error. One of the most important aspects of a multivariable controller is its ability to cope with this error. From a mathematical point of view, the controller inverts a matrix of the dynamic process model in order to calculate the MV moves. If the matrix contains constraints on highly interacting variables, it is poorly conditioned, and the solution is sensitive to model error. These are called ill-conditioned problems.

At any particular control execution, the process encountered by the controller may require excessive input movement in order to control the outputs independently.

It is important to note that this is a feature of the process to be controlled; any algorithm that attempts to control an ill-conditioned process must address this problem. A high condition number in the process gain matrix means that small changes in controller error will lead to large MV moves.

Two strategies are currently used by MPC controllers to accomplish this task—singular value thresholding and input move suppression. Readers are requested to refer to Camacho (2013) and Qin (2003) for further readings.

2.4.7   Step 6: Optimum Steady-State Targets

The next step in the MPC algorithm is to *calculate optimum steady-state targets* for all manipulated and controlled variables, as shown schematically in Figure 2.2. Almost all of the MPC products available perform a separate local steady-state optimization at each control cycle to compute steady-state input, state, or output targets. This is necessary because the optimal targets may change at any time step due to disturbances entering the loop or operator inputs that redefine the control problem. The steady-state linear program (LP) or quadratic program (QP) does this calculation.

The input to this calculation consists of the MV (manipulated variable) current values and operating limits, the CV (controlled variable) steady-state predicted values, economic information on values of products, and costs of raw materials and utilities.

The operating limits define an acceptable operating region and the MV current values and CV predicted steady-state values define the predicted steady-state operating point, assuming no moves in the MVs. This point may or may not be inside the acceptable operating region.

The optimization problem is typically formulated so as to drive steady-state MVs and CVs as closely as possible to targets determined by the local economic optimization, without violating input and output constraints. Constant disturbances estimated in the output feedback step appear explicitly in the steady-state optimization so that they can be removed.

The LP calculates a steady-state move in each MV, which, taken together, specify a steady-state operating point that is within the acceptable operating region. Further, this operating point is optimal from an economic standpoint. The distinguishing feature of an LP solution is that the optimal steady-state targets will lie at the vertex of the constraint boundary. In other words, this optimal steady-state operating point will always be at several limits.

Figure 2.8 shows an example of a system with two MVs (manipulated variables) and six CVs (controlled variables).

The MVs are reflux flow set point and reboiler steam flow set point. The CVs are overhead impurity, bottoms impurity, flooding, top valve opening, top temperature, and bottom level, as shown in Figure 2.4 and given in Table 2.1. The operating limits on all six variables define an acceptable operating region.

In this example, the current steady-state operating point is inside the region, although this is not always the case. Also, this current *steady-state* operating point is not the same

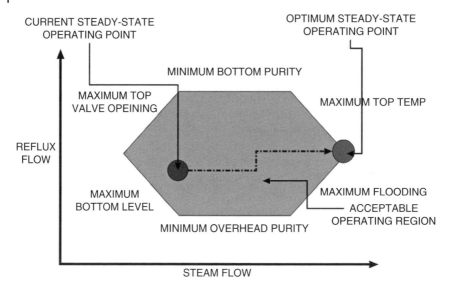

Figure 2.8 Operating region of a distillation column with two manipulated variables and six controlled variables

as the current operating point; it is the point to which the system is predicted to go in the absence of control action. Knowing the current steady-state operating point, the limits on all six variables, and economic information, it is possible to identify the optimal steady-state operating point. This point specifies optimal steady-state targets for all MVs and CVs.

Some MPC algorithms (like RMPCT) use quadratic programs for the steady-state target calculation. The QP solution does not necessarily lie at the constraint boundary, so the optimal steady-state targets tend not to bounce around as much as for an LP solution.

The MPC algorithm solves the local steady-state optimization problem using a sequence of LPs and/or QPs. CVs are ranked by priority such that control performance of a given CV will never be sacrificed in order to improve performance of a lower-priority CV. The prediction error can be spread across a set of CVs by grouping them together at the same priority level. The calculation proceeds by optimizing the highest-priority CVs first, subject to hard and soft output constraints on the same CVs and all MV hard constraints. Subsequent optimizations preserve the future trajectory of high-priority CVs through the use of equality constraints. Likewise, MVs can be ranked in priority order so that inputs are moved sequentially toward their optimal values when extra degrees of freedom permit.

2.4.8 Step 7: Develop Detailed Plan of MV Movement

The final step in the MPC algorithm is to *develop a detailed plan of control action* for the manipulated variables that minimizes the difference between the predicted future behavior and the desired future behavior of the controlled variables.

In other words, at the dynamic optimization level, an MPC controller must compute a set of MV adjustments that will drive the process to the desired steady-state operating

Figure 2.9 Revised CV trajectory and steady state error

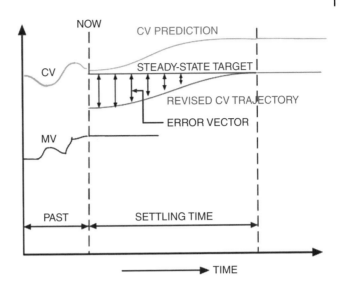

point without violating constraints. The MPC controller calculates a set of multivariable MV moves into the future to minimize the controller error.

Controller error is the difference between the target CV trajectory (SP) and the revised unforced CV response, as shown Figure 2.9.

The desired future behavior of the controlled variables is to have them at their steady-state targets, as calculated by the steady-state LP and/or QP.

Basically, the CV steady-state targets are the set points for the CVs. In order to dynamically drive the CVs to their targets, a series of future moves is calculated for each MV, extending approximately one-half steady-state time into the future. This allows the controller to defer control action, and also to better play the effects of one MV against the other MVs in solving the multivariable control problem.

The value of the MV, when all calculated moves are added in, must be equal to the MV steady-state target from the linear program. If all MVs reach their steady-state targets, the CVs will also reach their steady-state targets.

Figure 2.10 shows the development of a detailed control action for a one MV, one CV system. The desired effect of the control action is defined by the mirror image of the CV prediction about the CV steady-state target (CV set point).

If control action could be found that had exactly the desired effect, the error would be exactly canceled out. The CV would go immediately to the steady-state target and remain there across the future time horizon. This strategy is different in different vendor MPC algorithms, and more detail can be found in Camacho (2013) and Qin (2003).

Since we know the effect needed from the control action, and since the model of the process describes the effect on a dependent variable of a move in an independent variable, it is reasonable to assume that the model can be used in planning the future MV control moves.

Figure 2.10 also shows the control action planned for the MV. Figure 2.11 shows controlled variable predictions with and without control moves. Note that the CV plot displays a "CV Prediction with Control Moves." This curve is the result of adding the effect of the calculated control action to the CV prediction (based only on past moves

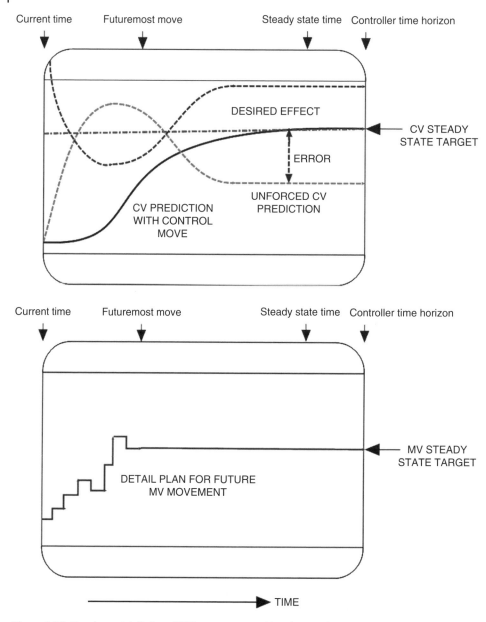

Figure 2.10 Develop a detail plan of MV movement to drive the steady state error to zero

in the independent variables). The unforced CV prediction curve is the CV prediction without any future MV movement.

Figures 2.11 and 2.12 show a snapshot of a controller solution for the distillation column shown in Figure 2.3. These figures will be used to demonstrate the key features of the MPC controller.

Each box contains information on one CV or MV. The y-axis is in engineering units of the CV or MV, while the x-axis is time. The left side of each box represents current time.

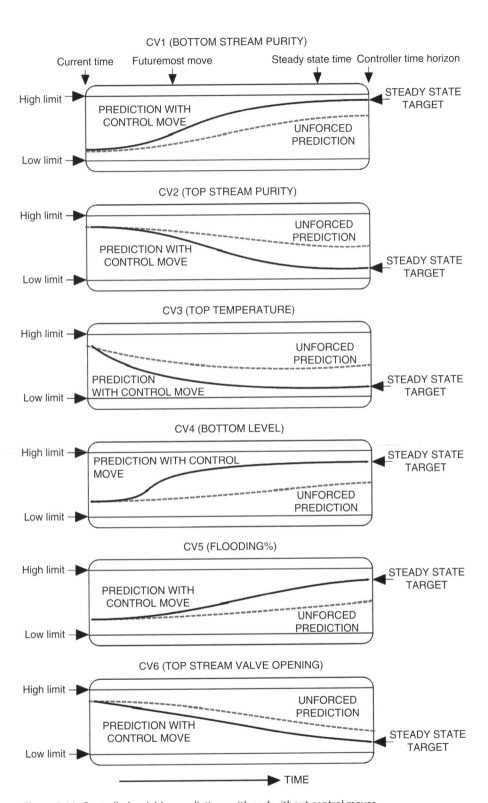

Figure 2.11 Controlled variables predictions with and without control moves

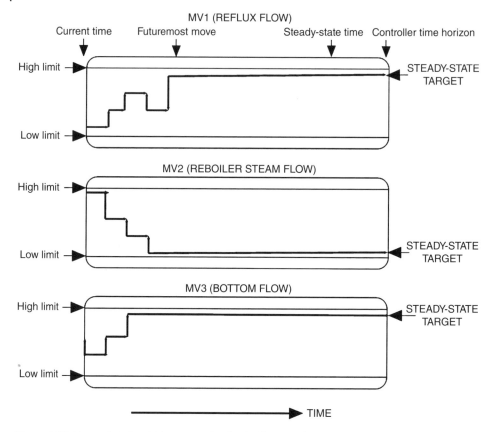

Figure 2.12 Manipulated variables move plan for distillation column

Notice that the future moves are calculated approximately halfway across the time to reach steady state.

Also notice that the right-hand side of each box is beyond the steady state time, referred to as the *controller time horizon*. This extension of time is required to allow the entire effect of the futuremost move to be seen. So the controller time horizon is equal to the steady-state time plus the time of the futuremost move.

Each of the six CVs has a prediction denoted by the dotted line. These predictions are based on the past history of the six independent variables and the model shown previously in this section (Figure 2.5).

These predictions represent where the six controlled variables are predicted to go in the absence of any control action.

Also note that all of the MVs and CVs have upper and lower operating limits. These limits define an acceptable operating region. The predicted steady-state values of the CVs and the current values of the MVs define the current steady-state operating point.

Using this point, economic information, and the acceptable operating region defined by the operating limits, the steady-state LP or QP calculates the optimal steady-state operating point. This appears as the steady-state target on the right side of each box. There is a steady-state target for every MV and CV in the controller.

The next step is to develop a detailed plan of control action. This plan can be seen in Figure 2.12, where a series of future moves has been calculated for each MV. These moves extend about halfway across the steady-state time and are required to reach the MV stead-state target.

These moves are calculated by minimizing the errors for all six CVs between the predictions and the CV steady-state targets. Figure 2.11 shows "Prediction with Control Moves" for each CV. This curve represents how that CV is predicted to respond in the future, *based on the control action* shown in Figure 2.12.

Notice that all MVs and CVs are predicted to end up at their respective steady-state targets, calculated by the steady-state LP/QP. This is no accident; if the MVs are required to end up at their steady-state target, the CVs *must* end up at their targets also.

A final point to make is that Figures 2.11 and 2.12 represent a single calculation of the controller at one control interval only. The first move of the 14 calculated in each MV is sent to the regulatory control system, and the rest of the moves are thrown away.

If, at the next control cycle, no disturbances have entered the system and model mismatch is negligible, the solution will be very similar to the remainder of the solution from the previous cycle.

However, if a major disturbance has entered the system, the optimal steady-state operating point will change, as will the predictions themselves (reflecting the disturbance). This will require an immediate change in control strategy, so the control action in this scenario will not resemble the solution from the previous control cycle.

It is for this reason that all the moves except for the first one are thrown away. It is still essential that these future moves be calculated, since the size of the first control move will be affected by projected MV constraint violations in the future. In other words, a particular MV might have to be moved more now in order to prevent a future limit violation. This would not be known if the entire trajectory of future moves was not calculated, subject to the operating limits.

References

Badgwell, T. A., & Qin, S. J. (2015). Model-Predictive Control in Practice. *Encyclopedia of Systems and Control*, 756–760.

Camacho, E. F., & Alba, C. B. (2013). *Model Predictive Control*. New York: Springer Science & Business Media.

Cutler, C., Morshedi, A., & Haydel, J. (1983). An industrial perspective on advanced control. In AICHE annual meeting. Washington, DC, October 1983.

Garcia, C. E., Prett, D. M., & Morari, M. (1989). Model predictive control: theory and practice—a survey. *Automatica*, 25(3), 335–348.

Kouvaritakis, B., & Cannon, M. (2016). *Model Predictive Control*. New York: Springer.

Morari, M., & Lee, J. H. (1999). Model predictive control: past, present and future. *Computers & Chemical Engineering*, 23(4), 667–682.

Qin, S. J., & Badgwell, T. A. (2003). A survey of industrial model predictive control technology. *Control Engineering Practice*, 11(7), 733–764.

Yousfi, C., & Tournier, R. (1991). Steady-state optimization inside model predictive control. In Proceedings of ACC'91, Boston, MA (pp. 1866–1870).

3

Historical Development of Different MPC Technology

3.1 History of MPC Technology

Model predictive control algorithm is at the heart of traditional advanced process control applications. Started during 1975, it has now dominated industrial process plant for last 40 years and brings billions of dollars of profits. MPC business reached their peak in late 1990s and early 2000s when the major licensors were caught up with going everywhere throughout the world and introducing model predictive controllers to bring benefits to its customers as quickly as time permitted. Qin provides an excellent survey of industrial model predictive control (Qin 2003; Qin 1997).

History and development of model predictive control can be found in Ruchika (2013). Today, existing applications are being revisited, modernized, and kept up, new applications are being introduced on new units and more open doors are being recognized on new areas and industries. Despite the fact that the foundation of all MPC calculations known today was created by very few yet exceptionally gifted groups, lately, there are not many licensors of MPC in the market, as the consequence of quite significant merging process of proven MPC technologies. All those algorithms are today in the hands of corporations, mostly vendors of control systems and instrumentation.

But how did it all begin? To understand the present, let's reveal some details from the past.

MPC technology started way back in 1975 for process industries. This is first-generation MPC technology (refer to Figure 3.1). After that, every 10 years there was a step jump of maturity in MPC technology. However, the theory of modern advance control can be traced back from 1960, where the first basic calculations were published by Kalman et al. (Kalman, 1960a, b). The following sections give a brief history of evolution of MPC technology.

3.1.1 Pre-Era

3.1.1.1 Developer

The pioneering work of modern control systems was done by Kalman et al. in the early 1960s (Kalman, 1960a, b). It is known as a linear quadratic Gaussian controller (LQG) or Kalman filter. In spite of the fact that Kalman methodology is considered to seed the fundamental establishment of MPC innovation, however, it does not cover every aspect of calculation, which we now call MPC innovation. That is the reason it is consider as pre-era of MPC innovation.

Multivariable Predictive Control: Applications in Industry, First Edition. Sandip Kumar Lahiri.
© 2017 John Wiley & Sons Ltd. Published 2017 by John Wiley & Sons Ltd.

3.1.1.2 Motivation

Basic motivation of Kalman was to develop an algorithm to control a system by solving a sequence of open-loop dynamic optimization.

Kalman filtering is an algorithm that uses a series of measurements observed over time, containing statistical noise and other inaccuracies, and produces estimates of unknown variables that tend to be more precise than those based on a single measurement alone, by using Bayesian inference and estimating a joint probability distribution over the variables for each timeframe. The filter is named after Rudolf E. Kalman, one of the primary developers of its theory.

3.1.1.3 Limitations

Being at a nascent stage of development, Kalman algorithm does not look into details of unresolved issues such as constraints, process nonlinearities, and robustness. As the algorithm was totally new and made by an academic researchers, industrial community at that time could not gauge its potential application and fail to accept it. However one major contribution of the Kalman algorithm was to boost interest to develop automated model predictive control application.

3.1.2 First Generation of MPC (1970–1980)

The Kalman filter boosted the interest to applying computer algorithm in process control. During the end of the 1970s, when the cost of the computer falls, process control practitioners in chemical industry start thinking about how to use process model to predict the future values of controlled variables and apply computer algorithms to control them within bounds. This concept was presented by two independent teams during the late 1970s: IDCOM and DMC algorithms are the first generation of MPC as we know it today.

Richalet et al. in a 1976 conference came up with first presentation of this breakthrough MPC technology. They described their approach as model predictive heuristic control (MPHC) and the solution software IDCOM, an acronym for Identification and Command. In today's context, the algorithm would be referred to as a linear MPC controller. Later, he summarized his research in a 1978 *Automatica* paper (Richalet et al., 1978). The first industrial use of this MPC technology was applied in fluid catalytic cracking units in refineries, with reported huge benefits.

Almost the same time, a team of engineers at Shell Oil led by Cutler and Ramaker presented their own MPC technology called Dynamic matrix control (DMC) at 1979 National AIChE meeting (Cutler & Ramaker, 1979) and at the 1980 Joint Automatic Control Conference (Cutler & Ramaker, 1980). Prett and Gillette (1980) described an application of DMC technology to FCCU reactor/regenerator control.

The initial IDCOM and DMC algorithms represent the first generation of MPC technology and had an enormous impact on industrial process control.

3.1.2.1 Characteristics of First-Generation MPC Technology

First-generation MPC technology introduced for the first time model-based control methodology in which the dynamic optimization problem is solved online at each control execution. Based on past history of manipulated variables movement, future plant behavior is predicted and optimized over a time interval known as *prediction horizon*. First-generation MPC technology has also developed step tests based on new process identification technology, in which quick estimation of empirical dynamic models from test data can be performed. This new identification technology reduces the cost and effort of model development and can make explicit process model easily, which can take, in principle, any required mathematical form. Process input and output constraints are included directly in the problem formulation so that future constraint violations are anticipated and prevented. This new methodology for industrial process modeling and control is what we now refer to as MPC technology.

3.1.2.2 IDCOM Algorithm and Its Features

The distinguishing features of the IDCOM approach are:

- Impulse response model for the plant, linear in inputs or internal variables
- Quadratic performance objective over a finite prediction horizon
- Future plant output behavior specified by a reference trajectory
- Input and output constraints included in the formulation
- Optimal inputs computed using a heuristic iterative algorithm, interpreted as the dual of identification

The algorithm looks into a process as an input and output black-box type model. Process inputs are divided into manipulated variables (MVs), which the controller adjusts, and disturbance variables (DVs), which are not available for control. Process outputs are referred to as controlled variables (CVs). They chose to describe the relationship between process inputs and outputs using a discrete-time finite impulse response (FIR) model. For the single-input, single-output (SISO) case, the FIR model looks like:

$$y_{k+j} = \sum_{i=1}^{N} h_i u_{k+j-i}$$

This model predicts the values of future control variables (y) by a linear combination of past manipulated values (u) and impulse response coefficients (h). The sum is truncated at the point where past inputs no longer influence the output; this representation is therefore only possible for stable plants.

The finite impulse response coefficients were identified from plant test data by fitting a model curve with actual step test data. Because the control law is not linear and could not be expressed as a transfer function, Richalet et al. refer to it as heuristic. In today's context, the algorithm would be referred to as a linear MPC controller.

The MPHC algorithm drives the predicted future output trajectory as closely as possible to a reference trajectory, defined as a first-order path from the current output value to the desired set point. The speed of the desired closed-loop response is set by the time constant of the reference trajectory. This is important in practice because it provides a natural way to control the aggressiveness of the algorithm. Increasing the time constant leads to a slower but more robust controller.

Richalet et al. visualize at that time the importance of hierarchical structure of process control. They describe four levels of control:

Level 3—Time and space scheduling of production
Level 2—Optimization of set points to minimize costs and ensure quality and quantity of production

Level 1—Dynamic multivariable control of the plant
Level 0—Base-level PID control

They point out that significant benefits do not come from simply reducing the variations of a controlled variable through better dynamic control at level 1. The real economic benefits come at level 2, where better dynamic control allows the controlled variable set point to be moved closer to a constraint without violating it. This argument provides the basic economic motivation for using MPC technology. The concept of a hierarchy of control functions is fundamental to advanced control applications and seems to have been followed by many practitioners.

3.1.2.3 DMC Algorithm and Its Features

Key features of the DMC control algorithm include:

- Linear step response model for the plant
- Future plant output behavior specified by trying to follow the set point as closely as possible
- Optimal inputs computed as the solution to a least squares problem

The linear step response model used by the DMC algorithm relates changes in a process output to a weighted sum of past input changes, referred to as input moves. For the SISO case, the step response model looks like:

$$y_{k+j} = \sum_{i=1}^{N-1} S_i \Delta u_{k+j-i} + S_N \Delta u_{k+j-N}$$

The move weights are the step response coefficients. A detailed description of the algorithm is discussed in Chapter 16.

The objective of a DMC controller is to drive the output as close to the set point as possible in a least squares sense with a penalty term on the MV moves. DMC first introduced a move-suppression concept of MVs to provide a more robust control. This results in smaller computed input moves and a less aggressive output response. As with the IDCOM reference trajectory, this technique provides a degree of robustness to model error. Move suppression factors also provide an important numerical benefit in that they can be used to directly improve the conditioning of the numerical solution.

The initial IDCOM and DMC algorithms represent the first generation of MPC technology; they had an enormous impact on industrial process control and served to define the industrial MPC paradigm.

3.1.3 Second-Generation MPC (1980–1985)

The original IDCOM and DMC algorithms provided excellent control of unconstrained multivariable processes. However, due to its nascent stage of development, they could not give light on constraint handling in real plants. Engineers at Shell Oil continued to develop the MPC algorithm and addressed this weakness by injecting quadratic program (QP) in DMC algorithm in which input and output constraints appear explicitly. Cutler et al. first described the QDMC algorithm in a 1983 AIChE conference paper (Cutler, Morshedi, & Haydel, 1983). Garcia and Morshedi (1986) later published a detailed algorithm of this second-generation technology. Due to their enormous contribution to develop first- and second-generation MPC algorithms, Cutler and Ramaker are considered as the fathers of MPC technology.

The QDMC algorithm can be regarded as representing a second generation of MPC technology, comprising algorithms that provide a systematic way to implement input and output constraints. This was accomplished by posing the MPC problem as a QP, with the solution provided by standard QP codes.

Key features of the QDMC algorithm include:

- Linear step response model for the plant
- Quadratic performance objective over a finite prediction horizon
- Future plant output behavior specified by trying to follow the set point as closely as possible subject to a move suppression term
- Optimal inputs computed as the solution to a quadratic program

Garcia and Morshedi show how the DMC objective functions can be rewritten in the form of a standard QP. Future projected outputs can be related directly back to the input move vector through the dynamic matrix; this allows all input and output constraints to be collected into a matrix inequality involving the input move vector. Although the QDMC algorithm is a somewhat advanced control algorithm, the QP itself is one of the simplest possible optimization problems that could be posed. One big advantage of this algorithm is that a solution can be found readily using standard commercial QP optimization codes.

Garcia and Morshedi published their findings regarding industrial application in a pyrolysis furnace. The QDMC controller adjusted fuel gas pressure in three burners in order to control stream temperature at three locations in the furnace. Their test results demonstrated dynamic enforcement of input constraints and decoupling of the temperature dynamics. They reported good results on many applications within Shell on problems as large as 12×12 (12 process outputs and 12 process inputs). They stated that above all, the QDMC algorithm had proven particularly profitable in an online optimization environment, providing a smooth transition from one constrained operating point to another.

3.1.4 Third-Generation MPC (1985–1990)

The IDCOM-M, HIECON, SMCA, and SMOC algorithms represent a third generation of MPC technology; others include the PCT algorithm sold by Profimatics and the RMPC algorithm sold by Honeywell (refer to Figure 3.1).

From 1970 to 1985, MPC technology was in its developmental phase, and a few big companies (like Shell oil) were internally developing and utilizing it and getting huge profits. Other, smaller companies could not afford to develop their own technology. However, after initial phase, MPC technology slowly started to get huge profit to gain wider acceptance during the 1990s. During the 1980s, MPC algorithms were struggling to tackle larger and more complex problems of industry. Engineering teams continued the development of MPC algorithms and brought new implementations for improved handling. All groups were focusing on how to improve handling of the constraints, fault tolerance, objective functions, and degrees of freedom in their algorithms.

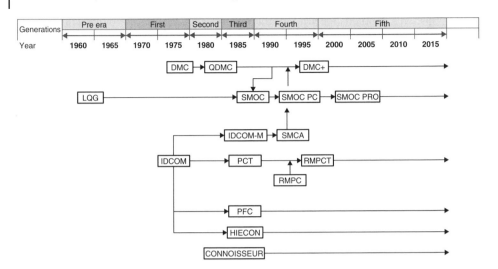

Figure 3.1 Brief history of development of MPC technology

As the result of these improvements, the vendors presented upgraded technologies, which were applied widely in industry during the 1990s:

- Setpoint presented improved IDCOM by the name IDCOM-M, which was later offered as SMCA (Single Multivariable Control Architecture)
- DMC group was separated from Shell Oil and developed improved DMC algorithm by the name of QDMC and was later bought by AspenTech
- Shell Oil continued their work and developed SMOC (Shell Multivariable Optimizing Controller)
- Adersa presented nearly identical algorithm: HIECON (Hierarchical constraint control)
- Profimatics had their PCT algorithm
- Honeywell with RMPC

Those are the algorithms, which represent MPC technology generation of the 1990s.

3.1.4.1 Distinguishing Features of Third-Generation MPC Algorithm

As MPC technology gained wider acceptance, and problems tackled by MPC technology grew larger and more complex, control engineers implementing second-generation MPC technology in real plants ran into many practical problems. Third-generation MPC technology addressed those problems in real plants and develop superior technology.

The main distinguishing features of third-generation MPC technology are as follows:

- *Mechanism to recover from an infeasible solution:* The soft constraint formulation of second-generation MPC is not completely satisfactory because it means that all constraints will be violated to some extent, as determined by the relative weights. However, in real plants some constraints are extremely important in terms of plant safety or product quality and should never be violated. Third-generation MPC technology makes adjustments in the algorithm so that these important constraints are never violated and low-priority constraints are shed in order to satisfy higher-priority ones.

- *Mechanism to address the issues resulting from a control structure that changes in real time:* In practice, process inputs and outputs can be lost in real time due to signal hardware failure, valve saturation, or direct operator intervention. They can just as easily come back into the control problem at any sample interval. This means that the structure of the problem and the degrees of freedom available to the control can change dynamically. MPC algorithms were modified to calculate the size of control problems in each execution and address any issues resulting from a control structure that changes in real time. For a situation when number of MVs available is much lower than number of CVs, it will not be possible to meet all of the control objectives. The control specifications must be relaxed somehow—for example, by minimizing CV violations in a least-squared sense.

- *Fault tolerance capability:* Third-generation MPC are equipped with fault tolerance mechanism. Rather than simply turning itself off as signals are lost, a practical MPC should remain online and try to make the best of the subplant under its control. A major barrier to achieving this goal is that a well-conditioned multivariable plant may contain a number of poorly conditioned subplants. In practice, an MPC must recognize and screen out poorly conditioned subplants before they result in erratic control action.

The practical issues of second-generation MPC technology motivated engineers at Adersa, Setpoint, Inc., and Shell (France) to develop third generations of MPC technology. The version marketed by Setpoint was called IDCOM-M (the M was to distinguish this from a SISO version called IDCOM-S), while the nearly identical Adersa version was referred to as hierarchical constraint control (HIECON). The IDCOM-M controller was first described in a paper by Grosdidier, Froisy, and Hammann (1988). A second paper presented at the 1990 AIChE conference describes an application of IDCOM-M to the Shell Fundamental Control Problem (Froisy & Matsko, 1990) and provides additional details concerning the constraint methodology.

3.1.4.2 Distinguishing Features of the IDCOM-M Algorithm

Distinguishing features of the IDCOM-M algorithm include:

- Linear impulse response model of plant
- Controllability supervisor to screen out ill-conditioned plant subsets
- Multiobjective function formulation
- Quadratic output objective followed by a quadratic input objective
- Controls a subset of future points in time for each output, called the coincidence points, chosen from a reference trajectory
- A single move computed for each input
- Either hard or soft constraints, with hard constraints ranked in order of priority

An important distinction of the IDCOM-M algorithm is that it uses two separate objective functions, one for the outputs and then, if there are extra degrees of freedom, one for the inputs. A quadratic output objective function is minimized first, subject to hard input constraints.

Setpoint engineers continued to improve the IDCOM-M technology, and eventually combined their identification, simulation, configuration, and control products into a single integrated offering called SMCA, for Setpoint Multivariable Control Architecture. An improved numerical solution engine allowed them to solve a sequence of separate steady-state target optimizations, providing a natural way to incorporate multiple ranked control objectives and constraints.

3.1.4.3 Evolution of SMOC

In the late 1980s, engineers at Shell Research in France developed the Shell Multivariable Optimizing Controller (SMOC) (Marquis & Broustail, 1998; Yousfi & Tournier, 1991), which they described as a bridge between state-space and MPC algorithms. The distinguishing feature of SMOC algorithm is that it combines the constraint handling features of MPC with the richer framework for feedback offered by state-space methods.

3.1.4.4 Distinctive Features of SMOC

The SMOC algorithm includes several features that are now considered essential to a modern MPC formulation:

- State-space models are used so that the full range of linear dynamics can be represented (stable, unstable, an explicit disturbance model describes the effect of unmeasured disturbances); the constant output disturbance is simply a special case.
- A Kalman filter is used to estimate the plant states and unmeasured disturbances from output measurements.
- A distinction is introduced between controlled variables appearing in the control objective and feedback variables that are used for state estimation.
- Input and output constraints are enforced via a QP formulation.

Details of SMOC evolution and its technical features are discussed in Chapter 16.

3.1.5 Fourth-Generation MPC (1990–2000)

After the nascent stage of 1970 to 1980, MPC business increased and reached its peak in the late 1990s. In parallel with the market growth, the competition between the vendors was also reaching its peak during late 1990s. Major MPC companies started making acquisitions and wanted to dominant the market. AspenTech and Honeywell got out as the winners of this phase.

In the era of 1990–2000, increased competition and the mergers of several MPC vendors led to significant changes in the industrial MPC landscape. In late 1995, Honeywell purchased Profimatics, Inc. and formed Honeywell Hi-Spec Solutions. The RMPC algorithm offered by Honeywell was merged with the Profimatics PCT controller to create their current offering called RMPCT. In early 1996, Aspen Technology Inc. purchased both Setpoint, Inc. and DMC Corporation. This was followed by acquisition of Treiber Controls in 1998. The SMCA and DMC technologies were subsequently merged to create Aspen Technology's current DMC-plus product. DMC-plus and RMPCT are representative of fourth-generation MPC technology.

APC applications present on the market during the 2000s were:

- AspenTech
- Honeywell
- APCS (Adaptive Predictive Control System): SCAP Europa
- DeltaV MPC: Emerson Process Management

- SMOC (Shell Multivariable Optimizing Controller): Shell
- Connoisseur (Control and Identification package): Invensys
- MVC (Multivariate Control): Continental Controls Inc.
- NOVA-NLC (NOVA Nonlinear Controller): DOT Products
- Process Perfecter: Pavilion Technologies

To support the market needs, some of these technologies started to include nonlinear MPC, while most of them were still linear MPC.

3.1.5.1 Distinctive Features of Fourth-Generation MPC
- Windows-based graphical user interfaces
- Multiple optimization levels to address prioritized control objectives
- Additional flexibility in the steady-state target optimization, including QP and economic objectives
- Direct consideration of model uncertainty (robust control design)
- Improved identification technology based on prediction error method and sub-space ID methods

These and other MPC algorithms currently available in the marketplace are described in greater detail in the Chapter 16.

3.1.6 Fifth-Generation MPC (2000–2015)

The consolidation of vendors continued over the starting years of the centuries as well. AspenTech and Honeywell, however, still managed to preserve their leading role on the Advance Process Control (APC) market. Today, we are witnessing a further technology development that is not so much focused on improving the algorithms, but on improving the development steps. The focus is to make those steps smoother, faster, and easier, both for the developer and for the client, and to do as much as possible remotely. A great amount of knowledge for online optimization has been gained over the decades, so the systems used for the development today are also smarter. However, shortening of time and resources during the development phases increases the risk to not define the "optimal" model and strategy.

Aspen Technology has introduced a set of innovative technologies that are having a significant impact on APC methodology, applications development, benefits, and sustainability. A traditional Aspen DMCplus® application pushes for full LP optimization that exploits all available degrees of flexibility. However, when a large mismatch exists between the plant and the controller model, the LP may exhibit what is called *flipping* behavior. This phenomenon drove elongated project cycles, as heavy plant testing was required in order to develop highly accurate models before the application could be deployed. This often resulted in many months elapsing before the application could be put online to start generating operational benefits.

The new technology, adaptive process control, is designed to solve these problems. First, adaptive process control maintains the LP steady-state target in an *optimum area* instead of an optimum point. Hence, optimization and stable behavior are achieved even in the presence of a very large model mismatch.

Second, APC can utilize what are called *seed* models that are built from pretest and/or purely historical process data or from an existing rigorous plant model. This means that—coupled with the robust optimization behavior in the presence of a model mismatch—the controller can be deployed with these less refined seed models, and as a result, in a much shorter period of time, begin accruing benefits while the model is improved online by the adaptive technology.

Third, once deployed, the acquisition of process data to refine the seed model is completed using small perturbation background step testing while the controller is online. The underlying enhancements to the model identification algorithm enable the effective use of significant amounts of closed-loop data that can be collected while a user-specified level of emphasis on optimization is carried out by the system. That specification enables the user with the ability to make trade-offs between the speed of developing updated models and the degree of benefits lost.

The collective set of innovations within APC is driving a radical change in the economics of APC applications. While it has been shown that the technology solves many of the maintenance issues with traditional APC solutions, it is also having a dramatic impact on the development of new APC applications.

Honeywell has developed similar applications. All these features are the earmark of fifth-generation MPC technology.

3.2 Points to Consider While Selecting an MPC

One critical question asked by industry personnel is, which MPC vendor is best? Which MPC technology is superior? The dilemma is which MPC vendor to choose to implement MPC application in a company's own plant.

There is no clear-cut answer to these questions. Technologically, all the MPC algorithms are equivalent. The basic algorithms are the same; they only differ cosmetically. It depends on individual plant own environment, cost of implementation, size and nature of plant, and benefits expected after MPC implementation.

Some of the criteria that should be investigated before choosing MPC vendors are as follows:

- Robustness of MPC technology.
- Cost of implementation of MPC application.
- Recurring cost of running and maintenance of MPC application.
- How easily a controller adapts itself to changing process conditions. This directly influences extent of intervention required.
- Capability of plant internal manpower to do periodic maintenance of MPC application. One aspect of MPC that is repeatedly ignored is the requirement to maintain it. It should be borne in mind that advanced controllers once they are designed and implemented also need to be maintained, just like any other piece of equipment such as a pump or compressor.
- How easily it can be maintained, model updated, or retuned.
- How easily it can be remodeled, if necessary, for gross excursions in process conditions from what it is designed for.
- How one validates a model developed.
- The peripherals that are required.
- How easily it can be integrated with existing control system.

- Expertise of vendor manpower. The reason can generally be pinned down to lack of personnel with adequate knowledge of control theories, unwieldy models, lack of clarity and transparency in model and controller design, and so on.
- How many controller vendors created in similar industries and their performance.

References

Cutler, C. R., & Ramaker, B. L. (1979). Dynamic matrix control—a computer control algorithm. AICHE national meeting, Houston, TX, April 1979.

Cutler, C. R., & Ramaker, B. L. (1980). Dynamic matrix control—a computer control algorithm. In Proceedings of the joint automatic control conference.

Cutler, C., Morshedi, A., & Haydel, J. (1983). An industrial perspective on advanced control. In AICHE annual meeting, Washington, DC, October 1983.

Froisy, J. B., & Matsko, T. (1990). IDCOM-M application to the Shell fundamental control problem. AICHE annual meeting, November 1990.

Garcia, C. E., & Morshedi, A. M. (1986). Quadratic programming solution of dynamic matrix control (QDMC). *Chemical Engineering Communications*, 46, 73–87.

Garcia, C. E., Prett, D. M., & Morari, M. (1989). Model predictive control: theory and practice—a survey. *Automatica*, 25(3), 335–348.

Grosdidier, P., Froisy, B., & Hammann, M. (1988). The IDCOM-M controller. In T. J. McAvoy, Y. Arkun, & E. Zafiriou (Eds.), *Proceedings of the 1988 IFAC workshop on model based process control* (pp. 31–36). Oxford: Pergamon Press

Kalman, R. E. (1960a). Contributions to the theory of optimal control. *Bulletin de la Societe Mathematique de Mexicana*, 5, 102–119.

Kalman, R. E. (1960b). A new approach to linear filtering and prediction problems. *Transactions of ASME, Journal of Basic Engineering*, 87, 35–45.

Kwakernaak, H., & Sivan, R. (1972). *Linear Optimal Control Systems*. New York: John Wiley & Sons.

Marquis, P., & Broustail, J. P. (1998). SMOC, a bridge between state space and model predictive controllers: Application to the automation of a hydro treating unit. In T. J. McAvoy, Y. Arkun, & E. Zafiriou (Eds.), *Proceedings of the 1988 IFAC workshop on model based process control* (pp. 37–43). Oxford: Pergamon Press.

Morari, M., & Lee, J. H. (1999). Model predictive control: past, present and future. *Computers & Chemical Engineering*, 23(4), 667–682.

Prett, D. M., & Garcia, C. E. (1988). *Fundamental Process Control*. Boston: Butterworths.

Prett, D. M., & Gillette, R. D. (1980). Optimization and constrained multivariable control of a catalytic cracking unit. In Proceedings of the joint automatic control conference.

Qin, S. J., & Badgwell, T. A. (1997, June). An overview of industrial model predictive control technology. In AIChE Symposium Series (Vol. 93, No. 316, pp. 232-256). New York, NY: American Institute of Chemical Engineers, 1971–2002.

Qin, S. J., & Badgwell, T. A. (2003). A survey of industrial model predictive control technology. *Control Engineering Practice*, 11(7), 733–764.

Rawlings, J. B., & Muske, K. R. (1993). Stability of constrained receding horizon control. *IEEE Transactions on Automatic Control*, 38(10), 1512–1516.

Richalet, J., Rault, A., Testud, J. L., & Papon, J. (1976). Algorithmic control of industrial processes. In Proceedings of the 4th IFAC symposium on identification and system parameter estimation. (pp. 1119–1167).

Richalet, J., Rault, A., Testud, J. L., & Papon, J. (1978). Model predictive heuristic control: Applications to industrial processes. *Automatica*, 14, 413–428.

Ruchika, N. R. (2013). Model predictive control: History and development. *International Journal of Engineering Trends and Technology (IJETT)*, 4(6), 2600–2602.

Scokaert, P. O. M., & Rawlings, J. B. (1998). Constrained linear quadratic regulation. *IEEE Transactions on Automatic Control*, 43(8), 1163–1169.

Yousfi, C., & Tournier, R. (1991). Steady-state optimization inside model predictive control In Proceedings of ACC'91, Boston, MA (pp. 1866–1870).

4

MPC Implementation Steps

4.1 Implementing a MPC Controller

Once a process unit is identified as a potential MPC application, then the following control project steps are required to implement the controller. Various MPC vendors described these steps in great detail in their product literature and training materials (see DMC Corp. 1994; Honeywell Inc. 1995; Muske 1993). The project steps are summarized in Figure 4.1.

In the following sections, each of these steps is discussed.

4.1.1 Step 1: Preliminary Cost–Benefit Analysis

Implementation of MPC in commercial plant requires investment. Like any other investment, the top management of any company desires to know the payback period of MPC project before taking investment decision. A scientific cost–benefit analysis is thus a necessity before implementation to accurately estimate the payback period, return on investment (ROI), and so on. Accurate estimation of payback period or ROI will help to convince plant management in their investment decision, which led to their direct and indirect support to implementation team. Management support for MPC implementation is necessary to deal with various undesirable conflicting issues in later stages of the project. The purpose of this step is to utilize scientific method for cost–benefit analysis of MPC project before implementation.

4.1.2 Step 2: Assessment of Base Control Loops

The regulatory PID control loop forms the base layer of control system. MPC works above this foundation layer as a supervisory controller. It is utmost important to assess the performance of base control layer and improve upon it before trying to build MPC application over it. In this step an overall assessment of base control layer was performed scientifically and corrective steps are taken to rectify limitations, if any. This essentially means identification of any problems in control valves (like hysteresis, stiction, valve oversize or undersize phenomena), measuring sensors (like noise, range of instruments, calibration etc.), PID controller tuning, oscillation in process parameters and apply various techniques like controller tuning, maintenance of control valves, and calibration of instruments to rectify the problems.

Multivariable Predictive Control: Applications in Industry, First Edition. Sandip Kumar Lahiri.
© 2017 John Wiley & Sons Ltd. Published 2017 by John Wiley & Sons Ltd.

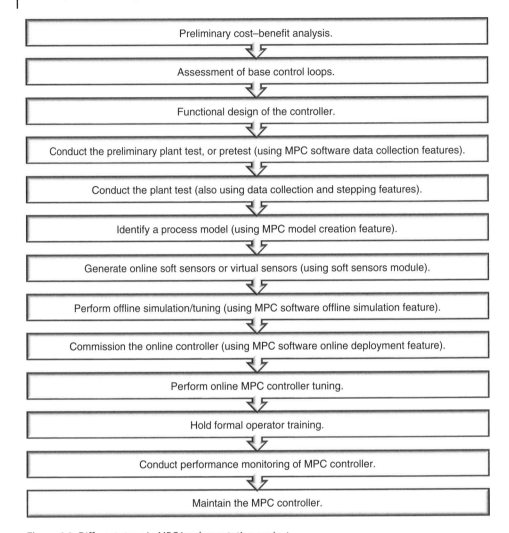

| Preliminary cost–benefit analysis. |
| Assessment of base control loops. |
| Functional design of the controller. |
| Conduct the preliminary plant test, or pretest (using MPC software data collection features). |
| Conduct the plant test (also using data collection and stepping features). |
| Identify a process model (using MPC model creation feature). |
| Generate online soft sensors or virtual sensors (using soft sensors module). |
| Perform offline simulation/tuning (using MPC software offline simulation feature). |
| Commission the online controller (using MPC software online deployment feature). |
| Perform online MPC controller tuning. |
| Hold formal operator training. |
| Conduct performance monitoring of MPC controller. |
| Maintain the MPC controller. |

Figure 4.1 Different steps in MPC implementation project

4.1.3 Step 3: Functional Design of Controller

Before building MPC application for a process, the control engineer must understand all the relevant aspects of the process, its various limitations, how it makes profit, and what areas can be exploited to increase profit. As a starting step, MPC engineers usually survey PFD and P&IDs of the process under study, meet with operations and specialized workforce, and detail a preliminary controller outline. In this step, MPC engineers usually study the control aspects of the process, analyze historical data, and understand the various safety and process constraints and equipment limitations. In this step, MPC engineers formulate the MV-CV-DV lists, understand what are the limits and constraints of the plant, and identify where there is an opportunity to gain more profit and how to exploit those margins available through the MPC application. The success of the MPC controller will greatly depend on how the functional designs are formulated to tap the potential margins available in the process. This step requires synergy between expertise and experience of domain engineer or plant process engineers and MPC engineers.

4.1.4   Step 4: Conduct the Preliminary Plant Test (Pre-Stepping)

This step is conducted just before the plant step test, where several step changes are given in MVs in series and data are collected to see its effect on different CVs with time. Preliminary plant tests can be seen as a preparatory step of the main step test.

There may be many obstacles/disturbances to doing a step test. Obstacles include data collection system not performing well and step sizes being too small to make a substantial change in relevant CVs above its noise level. The idea of pre-stepping is to identify all these unforeseen situations in the early stages and rectify them beforehand. The basic objective of pre-stepping is to save time in the main step test. Another key objective of pre-stepping is to identify the size of moves required in the MVs and DVs to observe a change (above the noise level) in the CVs. In this step, settling time of the system is also determined. MPC software data collection feature can be used to collect and extract preliminary plant test data.

4.1.5   Step 5: Conduct the Plant Step Test

Once base regulatory controller performance has been evaluated and improved and all the issues identified in preliminary plant step have been resolved, the *plant step test* can be performed.

This is done by performing a series of step moves (perturbations) in each independent variable and collecting the process data during these moves. The step test can be done manually or automatic (PRBS). Depending on plant size and number of control loops, step test keeps going anyplace from a couple days to a few weeks. Control engineering personnel work with the operators in the control room around the clock for the duration of the test. A preliminary process model is also done in parallel during step test data collection to ensure good data quality. This plant step test is absolutely the *most critical and most time-consuming* step in a control project. As the models used in MPC are data driven, those models are as good or bad as the quality of step test data. With accurate models, the time spent commissioning the controller will be greatly reduced, assuring the long-term success of the controller. If, however, the test is not done correctly, the commissioning will not proceed smoothly, and in all likelihood, the test will have to be repeated.

4.1.6   Step 6: Identify a Process Model

In the next step, *model identification* is performed using the process data recorded in the plant step test. This essentially means developing a MV-CV relationship for each pair. Data collected during the plant test are moved to an offline PC and with the help of MPC, model-building software process models—that is, MV-CV relationships—are generated. This step usually involves inspection and analysis of step test data, generation of more than one model curve for each MV-CV pair assuming different settling time, and utilization of different model equations and techniques available in MPC software to build the process model.

4.1.7 Step 7: Generate Online Soft Sensors or Virtual Sensors

When it is very difficult or costly to measure an important parameter online such as distillation tower top product impurity, soft sensors are used to predict that inferential property from other easy measurable parameters like top temperature, pressure, and so on. Sometimes soft sensors are used as back up to an existing analyzer to reduce or eliminate dead time, both from the process and the analyzer cycle.

This step involves identification of potential areas where a use of soft sensors can greatly enhanced the performance of MPC controllers and then developing robust and accurate soft sensors for it.

In this step, a scientifically step-by-step procedure is used to develop online soft sensors.

4.1.8 Step 8: Perform Offline Controller Simulation/Tuning

Before commissioning the controller in actual process plant, it is extremely important to know how the controller performs in real-time scenario but in offline mode. Offline simulation refers to running the controller in a separate offline PC to see the MV-CV dynamic responses of the process.

In this step MPC engineer build the controller configuration file using MPC software, then performing *offline simulation and tuning* of the controller using MPC offline simulation features. In this step, control engineers can test changes in disturbance variables, CV set points, LP costs, MV and CV limits, and so on and evaluate how controller performs in dynamic environment.

It is giving an opportunity to MPC engineers to assess the real-time performance of the developed MPC application. In offline simulations, different high–low limits of MV and CV are set with consultation with operation department. Different tuning parameters and LP costs of the controller are also set to make it effective on its real-time performance. After evaluating the controller performance, control engineers can make tuning changes to provide the desired dynamic performance of the controller. Control engineer then store the controller configuration file with its new values. The controller configuration file and its associated model file are then ready for the next step, online commissioning.

4.1.9 Step 9: Commission the Online Controller

At this point, *online controller commissioning* is performed. Commissioning of the controller means connecting the MPC controller online with the plant DCS and allow it to take control of the plant. The MPC online control program is first run in *prediction mode* for a day or two to verify proper program functioning, and check the accuracy of the model. In this mode, the controller performs all control calculations, but the moves are not sent to the process.

During this time, informal operator training takes place. A brief operator guide is provided, specifying the objective of the MPC controller, the variables that it considers, the operator displays to be used, and the procedure for turning the controller on and off.

The control engineer works with the operators to explain how to interpret the information on the operator displays and how to use the displays to interact with the MPC controller.

Once these steps are accomplished, the actual commissioning takes place. All manipulated variable limits are pinched in very near the current set point values so that only very small control moves are allowed.

The MPC controller is then turned on. Once the controller runs, calculates moves, and implements these moves on the process, the control engineer must make sure that the calculated moves are actually implemented at the regulatory control level.

At this point, the manipulated variable limits are gradually relaxed, controller performance is evaluated, and retuning is done as required. Commissioning of controllers at online platform in a running plant is a critical job and need 24 hours coverage. Care should be taken so that any mistake during commissioning should not lead to plant shutdown or plant upset. Most common mistake is to read and write to a wrong tag in DCS, induce process fluctuation due to MPC poor model or tuning.

4.1.10   Step 10: Online MPC Controller Tuning

When an MPC application is put online, it may not perform well as its performing in offline simulations mode. Reasons are many. This may be due to unmeasured disturbances that impact the CV too much, which the model cannot capture. Another probable reason may be that the step test model prediction is poor due to error in the model. Performance of MPC might not be up to par due to either process behavior or controller behavior. Purpose of online tuning is to tweak various tuning parameters of MPC controller, so that controller performance can be enhanced. In this step, controller dynamic performance is evaluated. Troubleshooting and proper tuning are done so that desired controller performance can be achieved.

4.1.11   Step 11: Hold Formal Operator Training

Operators and plant engineers are the ultimate end users of MPC application. It is absolutely necessary to get their support and confidence at every step of MPC implementation. If they are unhappy about the controller, it will be switched off. One of the key factors for success of MPC is a well-trained operator. Once the MPC controller has been successfully commissioned, *formal operator training* takes place. Delaying formal operator training until after commissioning allows the training to be done in the context of the final controller. At this point, operators and engineers are trained with MPC software front end and how to use it. A detailed training is also given to explain different features of MPC software, how it works, how to monitor its performance, and what to do if performance deteriorates.

4.1.12   Step 12: Performance Monitoring of MPC Controller

Like any other investments in process industry, MPC applications also need postimplementation payback calculations and performance assessment. Payback calculations are needed to convince management that putting money into MPC application is a profitable investment and can be considered as a top and attractive investment strategy. There are two types of performance assessment: First, one is to assess the MPC performance and calculate the benefits just after MPC implementation to justify the investment. Second, one is to periodically monitor MPC performance so that any deterioration in performance can be quickly detected and resolved.

A periodical performance review will also provide an idea of how much of initial benefits are preserved over time and how much money is lost by not getting the full potential performance. This will help to justify periodical maintenance or overhaul MPC application. If performance is deteriorating, then from the trend of performance KPI it can be decided when to do maintenance or overhaul MPC application.

Usually, MPC implementation projects end with this step. In most cases, the MPC project is done by external MPC vendors. In this step, MPC vendors hand over the project to the client, make a conclusion presentation to client plant management, and hand over all related documents to the client plant. However, to sustain the benefit of the project, it is absolutely necessary to monitor the performance of MPCs on a periodical basis and conduct preventive and reactive maintenance to enhance the performance. That constitutes the next step.

4.1.13   Step 13: Maintain the MPC Controller

For any MPC application, some amount of *maintenance* is required. The most frequently performed maintenance task is monitoring the operating limits on the manipulated and controlled variables in the problem, to be sure that they are spread out to reflect the largest operating region allowed. Left unattended, these limits often get pinched closer and closer, thus limiting the controller's ability to optimize the process. The other job is to periodically tune MPC online parameters so that any changes in the process can be dealt effectively.

There are many reasons why MPC controller fails. The main idea of this step is to understand those reasons and safeguard against them so that initial benefits of the controller can be sustained.

4.2   Summary of Steps Involved in MPC Projects with Vendor

In most of the cases, MPC projects are implemented by an external MPC vendor like Honeywell, Yokogawa, or Aspen Tech. Client plants usually receive help from these MPC vendors to implement MPC application in their plant as a turnkey project. Figure 4.2 briefly summarizes the different steps involved in such a turnkey project. The following section explains in brief the various steps depicted in Figure 4.2.

Client plants usually release a purchase order to MPC vendor to show their interest to implement MPC project. Upon receipt of purchase order, MPC vendor usually form a project team by assigning a project manager and experienced MPC engineers to the project. The team members familiarize themselves with project background material in preparation for a project kick-off meeting.

In the kick-off meeting, the teams from MPC vendor and the client decide on project organization, communication channel, and project schedule.

In order to assess the potential benefits of the project, the economic drives of the units under study is established. With knowledge of product economics, feed changes (if relevant), operational constraints, and plant historic data (lab results, plant data), the potential benefits are calculated.

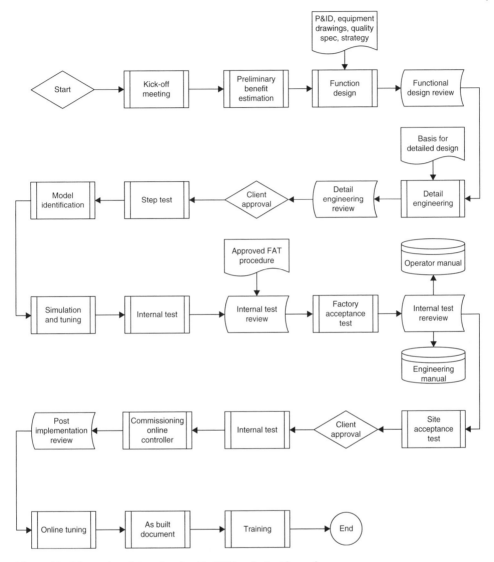

Figure 4.2 Schematics of steps involved in MPC project with vendor

Based on information gathered during base-level study, a preliminary basic functional design is made. The basic design is reviewed by the client. This study forms the basis for detailed engineering, which commences the next phase of the project.

During detailed design, activities like data configuration, interface design, details of hardware integration, procedures for step response tests, and are detailed and finalized. After the preliminary design review meeting, all comments on preliminary design are incorporated to prepare detailed design. This document usually submitted to client for approval. An approval of this document will constitute a design freeze of the project. Before going ahead with plant test, a pre-stepping step is performed to ensure that all instruments are in working order.

Once the client approves the detailed design, a 24-hour plant test is conducted. A high-quality plant test will generate an accurate model. It will enable control engineers to better understand process behavior and its constraints. The test response data are analyzed using a model identification software to generate a model.

The output at this stage is a process model, which is nothing but a dynamic relationship of manipulated variables with each controlled variable. The model so developed is run through an offline controller package. The controller behavior is studied for changes in set point, measured and unmeasured disturbances, and then is suitably tuned.

The next stage is to build online MPC controller using models developed. Observing the performance of controller in an environment that simulates process response, the controller is fine-tuned, relative priorities set for controlled and manipulated variables. The output is a tuned controller that can be read by MPC online software.

The next phase is factory acceptance tests to validate the system functionality. Prior to offering it to customer for tests, the system is tested thoroughly by MPC vendor's engineers. This test also presents a good opportunity for customer engineers to become acquainted with the system and train on it.

The performance of controller that has been developed will have to be tested at the site. This is the principal aim of the next step called a *site acceptance test*. Parameters of off-line controllers are downloaded and MPC is allowed to run on prediction mode. Prediction errors and calculated control moves are evaluated for accuracy. Once the dynamic model has been verified, controller is commissioned. Online performance is evaluated and adjustments are made as necessary.

After successfully commissioning the controller, the next and crucial step is to evaluate benefits accrued out of implementing MPC. This usually marks the end of project with MPC vendor.

References

DMC Corp. [DMC] (1994). Technology overview. Product literature from DMC Corp., July 1994.

Honeywell Inc. (1995). RMPCT concepts reference. Product literature from Honeywell, Inc., October 1995.

Muske, K. R., & Rawlings, J. B. (1993). Model predictive control with linear models. *AIChE Journal*, 39(2), 262–287.

5

Cost–Benefit Analysis of MPC before Implementation

5.1 Purpose of Cost–Benefit Analysis of MPC before Implementation

Implementation of MPC in a commercial plant requires investment. Like any other investment, top management of any company desires to know the payback period of MPC project before making investment decisions. Many MPC projects around the world have a payback period of six months or less. Traditionally, MPC has been applied in refining, petrochemical, and chemical industries and usually fetch huge benefits. Typical benefits are shown in Chapter 1.

Though MPC brings huge profit, nothing comes without a cost. MPC implementation costs as described as follows:

- Hardware, such as high-speed computer, wiring etc.
- Costs of MPC software and its license fees. These generally include:
  - MPC offline and online module
  - Data historian (like IP21, exaquantum etc.)
  - Offline soft sensor modeling software (like Aspen IQ)
  - Online postimplementation MPC performance monitoring software
- Cost of vendor manpower for MPC application development and implementation
- Cost of annual maintenance contract of MPC vendor support
- Cost of company's internal manpower to be involved during model building, step testing, deployment phase, postimplementation phase. This is a recurring investment.

All of these costs vary, depending on size and complexity of plant, number of controllers, vendor to vendor as per their scope assessment, and number of technical persons involved during MPC implementation stage from the vendor side.

Like any other investor, top management will always want to know where the return of this investment is. A scientific cost–benefit analysis is thus a necessity before implementation to accurately estimate the payback period and return on investment (ROI). A scientific and accurate estimation of ROI gives management the confidence to put money in MPC implementation and thus helps them to decide their priority to MPC as compare to other investment decisions that might be on their list.

Accurate estimation of ROI will help to convince them in their investment decision, which leads to their direct and indirect support of the implementation team. Management support is highly required for MPC implementation to deal with various

Multivariable Predictive Control: Applications in Industry, First Edition. Sandip Kumar Lahiri.
© 2017 John Wiley & Sons Ltd. Published 2017 by John Wiley & Sons Ltd.

undesirable conflicting issues in the later stage of implementation. Thus, it is necessary to know the scientific method for cost–benefit analysis before implementation. The next section deals with that.

5.2 Overview of Cost–Benefit Analysis Procedure

Reliable cost–benefit analysis start with the deep understanding of the following:

- How will MPC fetch benefits?
- What are the typical benefit percentages of already implemented MPC?
- For a new plant under study, what should be the scope of a scouting study?
- How should the scouting study be used to estimate benefits?

As described earlier, major benefits of MPC come from the following:

- Tighter control of key variables
- Increased production rates
- More stable and safe operations
- Increased plant flexibility
- Ability to move to optimum

Benefits may be calculated by the following methods:

- *Thumb rule method:* In this method, the MPC engineer calculates the benefits using an initial evaluation of plant, his past experience in similar plants, and some general rules of thumb. This method give ±30 percent accurate benefit values. This method can be used for initial estimates to see if a detail study is worthwhile.
- *Best running plant data method:* Usually, there is some point at which the plant runs most efficiently. In this method, the average values of key economic variables are compared to the historical best plant data values. The difference is the improvement possible with MPC. This method may underestimate benefits because MPC may outperform the best plant data.
- *Statistical method:* This method is often used in practice. Mean and standard deviations of key process variables (e.g., key quality and product yield parameters) are usually calculated. Reduction of standard deviation based on experience is used to estimate the benefit. Typical formula is:

$$\text{Savings} = \text{Value (\$/process effect)} \times \text{Gain} \times (\text{Standard deviation after}$$
$$\text{MPC} - \text{Standard deviation before MPC}) \times \text{Operating hours}$$
$$\text{per year} \times \text{Service factor of MPC}$$

where value is in $/Process effect and gain is the process effect such as selectivity and yield. A 50 percent reduction in standard deviation is often achievable; 60 to 100 percent reduction is achievable sometimes when the process dead time is small compared to disturbance frequency.
- *Step test method:* In this method, gains from step test curves are used to calculate benefits. The problem is that the actual step test has to be carried out to estimate the benefit, which is often not feasible at this stage.

- *Simulation method:* Detailed simulations can be carried out in commercial simulators to determine benefits from moving the process to a new operating zone. In most cases, good models are not available for complex industrial process (say polymer reactors, FCC, etc.). In practice, this method is rarely used currently but may be used more often in the future as more plants buy simulators for their processes.

To summarize, the statistical method is often the best technique, but it may not pick all the benefits. A more complex method may be used, but it is not worthwhile due to the fast payback times associated with advance control projects.

5.3 Detailed Benefit Estimation Procedures

Benefit estimation is a multidisciplinary act. It involves plant operators, production engineers, process engineers, control engineers, maintenance engineers, planning and scheduling departments, and plant heads having an idea of company vision and future path. The best way to understand the process and its economics is to perform a process analysis study with onsite data gathering. It is usually done by discussion and meeting with all relevant players at their site.

It includes the following steps:

1) Meet with operation, process engineering and all other relevant staff at their workplace (e.g., in control room, plant conference room etc.).
2) Explain the MPC project outline and objectives of this benefit estimation procedure.
3) Encourage the team to provide all relevant inputs. Collect their ideas.
4) Understand the potential areas where MPC can be exploited to reduce variation in key process variables and fetch benefits.

A detail benefits estimation procedure is shown in Figure 5.1.

5.3.1 Initial Screening for Suitability of Process to Implement MPC

In this step, it is decided whether the plant under study is a potential candidate for MPC implementation. To screen the plants for suitability of implementing MPC, it is necessary to know who the ideal candidates for MPC implementations are and what their characteristics are.

The primary candidates for MPC control are *high volume units*, such as crude distilling units in refinery and big petrochemical plants. Due to the large scale of operation of these units, a small improvement in operation results in a very significant economic benefit.

Conversion units, such as Cat Crackers and Hydrocrackers, are also good candidates for MPC. Since these units are converting a less valuable feed to higher-valued products, the potential exists to significantly improve yield and profitability.

A third group of candidates for MPC are *units that produce a highly valued product(s)*. Even if the unit is a low-volume unit, the benefits can be quite significant for increased product recovery. Many specialty chemicals applications fall under this group.

A fourth group of candidates are *units with a high consumption of utilities*. Many fractionation trains fall under this group. The products are overpurified so that the effects of disturbances do not cause the products to go off-specification. By implementing MPC,

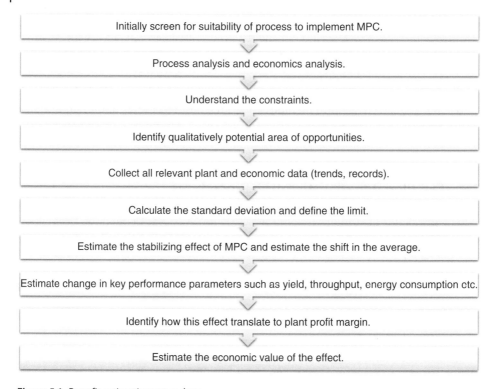

Figure 5.1 Benefit estimation procedure

the effect of the disturbances can be accounted for, with products controlled closer to specification, which saves energy.

A fifth group of candidates for MPC are *units that are subject to frequent disturbances.* The complex distillation towers in refinery provides a prime example of this type of application. Stabilizing the operation of the unit may not have tangible benefits at the point where the controller is implemented, but the profitability of the downstream units is generally improved. Constraints can be pushed more closely since the feed is more stable.

Based on these criteria, an initial screening may be done to know whether the plant under study qualifies for MPC implementation.

5.3.2 Process Analysis and Economics Analysis

A full understanding of the process is very much essential to estimate the potential benefit. There is no shortcut. There is no hard and fast rule on how to understand the process, but the following questions give an idea as to what to ask the team, which will facilitate the process:

- How will the unit or plant make a profit?
- What are the top management goals to operate this unit?
- How can profit be increased?
- What are the production targets? How are these targets met?

- What are the raw material and utility-specific consumption targets?
- What variables will indicate unit performance?
- Are there specific economic incentives to improve operating performance?
- What are the operating objectives of the unit?
- Is quality control currently good or poor?
- What problems do panel operators face on a daily basis?
- How many times does product quality go off spec in a month or year?
- How much loss does the plant incur due to an off-spec product?
- What areas of the plant consume the most time for panel operators?
- Which parameters must panel operators monitor most closely, and why?
- Is there any control problem or controller-tuning problem?
- How do other units of the refinery or complex affect the current process?

5.3.3   Understand the Constraints

Historically, MPC is expected to increase plant throughput between 2 and 10 percent. Look at the process with this viewpoint: Did plant capacity increase 2 to 10 percent after MPC implementation? Use these questions as a guideline:

- What are the constraints that prevent the plant from maximizing profit?
- Are there any raw material, utility limitations?
- Are there any capacity limitations in any major equipment like reactor, compressor, or distillation tower?
- How will catalyst efficiency, reaction selectivity/yield etc. be impacted by increase in capacity?
- What are the safety limits of different process parameters?
- What are the process limitations?
- What are the quality specs, and how does current plant quality stand against them?
- What are the process instrument limitations, such as valves at 100 percent open?

5.3.4   Identify Qualitatively Potential Area of Opportunities

This step is to identify the potential areas where MPC application can bring benefit after implementation. This is the area that needs experience and expertise. There are no hard and fast rules here; they vary from plant to plant, company to company. It is the job of the MPC engineer to quickly detect the potential profitable area while discussing with the plant operation and process engineering team. Note that some time data analysis and observing DCS parameter trends will help to pinpoint the potential areas.

The rule is to investigate the profitable areas and look for any opportunity that exists in the plant under investigation. It is helpful to keep in mind where MPC has been historically successful in capturing benefits. MPC engineers must investigate and discuss the presence of the following issues in the plant (not an exhaustive list):

- How will product distribution influence economics in a multiproduct distribution plant?
- Is there any area of the plant that exhibits strange or complex dynamics?
- Is there any area where the PID control loop is poorly controlling the process due to bad tuning or due to process interaction between process parameters?
- Is there any area with large dead time?

- Is there any area where there is large impact of ambient conditions? Is any equipment capacity limited by ambient conditions like condenser capacity due to high cooling water temperature in summer?
- Is there any scope to improve control by implementing a soft sensor? Is there any unmeasurable quality target?
- Currently, which equipment poses limits to increasing capacity further?
- How can MPC's stabilizing effect be used to increase plant throughput?
- Currently, how are product quality parameters controlled, and how far are they are from their respective limit of specifications?
- Is there any economic benefit to stabilizing the quality parameters?
- What is the frequency of plant producing off-spec product? What are the economic losses due to that? How can MPC's stabilizing effect be used to achieve less quality giveaway against specification?
- How can MPC be used to reduce time off spec during crude change or polymer grade change?
- Is there any selectivity, yield, or severity of operation gain by stabilizing the process?
- Which control areas require the most panel operator intervention and time?
- Is there any scope to reduce reflux and steam in any distillation column?
- Is there any scope to reduce flue gas in the furnace by accurately controlling the fuel oxygen ratio?
- Is there any scope to operate distillation column at low pressure and thus exploit the steam savings?
- How much is it feasible to utilize the surge vessel available in the plant (say, in distillation column bottom or overhead accumulator) to absorb the fluctuations and stabilize the downstream unit feed?
- What are the different limits (safety limit, process limit, interlock limit, quality spec limit, etc.) or constraints in the plant? How much margin is currently available in process or quality parameters that can be exploited by MPC?

Note that very detailed analysis is not possible at this stage of benefit estimation due to time constraints. Detailed analysis will be done in the functional design stage. Objectives of this benefit estimation step are to identify the major opportunity and quantify them. One major task is to identify the parameters whose data need to be collected for more accurate analysis and quantify them. A few examples are given next.

5.3.4.1   Example 1: Air Separation Plant

In an air separation plant, product oxygen is supplied to the downstream plants via multistage reciprocating compressor. Compressor discharge temperature remain at 180 °C with ±4 °C in summer where its trip limit was 185 °C. Further increase of oxygen flow could not be made due to trip point being so near. This trip point actually limits the oxygen supply and subsequently limits the capacity of downstream plant where this oxygen is used for reaction. The compressor has an interstage condenser operated by cooling water temperature. In winter, when cooling water temperature got reduced due to low ambient temperature, discharge temperature came down to 176 °C. Then there is a scope generated to increase the oxygen flow and return the discharge temperature to 180 °C again. Also, the difference between day and night ambient temperatures gives some scope to increase oxygen flow at night when there is margin in cooling water temperature. Also, MPC typically reduces the standard deviation to 50 percent. So it is expected to reduce the variation of discharge temperature from $\pm4°C$ to $\pm2°C$. Plant oxygen flow can be increased up to a point where discharge temperature can be operated at 182 °C instead of 180 °C after MPC implementation. In this way, throughput can be increased. Always attempt to determine if the potential area of plant capacity could not be increased due to ambient temperature or ambient conditions.

5.3.4.2 Example 2: Distillation Columns

Due to its stabilization effect and predictive capability, MPC is normally able to reduce reflux and steam. This subsequently reduces the flooding percentage in a distillation column and increases the margin to push more feed. So if a distillation column capacity is limited by flooding and entrainment, it is expected that MPC will fetch benefits. So look for the distillation columns in plants where such condition exists.

5.3.5 Collect All Relevant Plant and Economic Data (Trends, Records)

Collect all the historical data and trends for all the following parameters:

- Parameters identified in previous step
- All parameters that have influence on plant economics
- All parameters that represent any constraints
- All parameters that represent quality of product (online analyzer or offline laboratory analysis)
- Parameters where process control is not good, as per panel operators (These parameters have large swinging variations.)
- Parameters related to units where large dead time and complex process interactions exist

Normally, modern chemical plants that are considering implementing MPC are equipped with online data historians. It is good practice to collect historical data (say, for six months) of all the available tags (including offline laboratory analysis from laboratory information management systems (LIMS)) and later analyze what is required.

Typically, 15-minute average to hourly average data are collected to avoid low-frequency minute-to-minute noises.

5.3.6 Calculate the Standard Deviation and Define the Limit

Group the parameters unit-wise or process-wise and do a detailed study of all the parameters influencing economics and representing constraints or product quality, as identified in previous step. Evaluate the following:

- Clean the data for obvious errors such as instrument failure or plant shutdown.
- Find the standard deviation of each of the key process parameters and quality parameters. Do a complete statistical analysis of the data.
- From the data, find the average value or 50th percentile value or median and how far it is away from the spec or limit.

Case by case, fix the limit value of process parameters after discussing them with the process engineer. Some limits are obvious, like interlock value or quality spec. Some limits are not so obvious, like safety limit or maximum allowable furnace tube metal temperature.

5.3.7 Estimate the Stabilizing Effect of MPC and Shift in the Average

Industrial data contain noise due to many reasons. One must be careful to ensure that high-frequency noise in the data is not misread as control opportunity. Normally, MPC contributes by eliminating low-frequency disturbance in order of several minutes to an hour, in yield and throughputs. Therefore, to estimate potential yield shifts or throughput increases, one must look at longer-term variations rather than minute-to-minute variations.

The benefit estimation done by common statistical method is schematically represented in Figure 5.2. In general, any process parameter variation can be represented by its average value and standard deviation. MPC typical reduces the standard deviation by 50 percent and thus makes it possible for the operator to move the average value closer to its fixed limit, by shifting the set value, while the risks of temporary violations are constant or reduced. The trend in Figure 5.2 shows the variations of an important process variable over a period of time against its limit or constraint. The process variable can be product quality, any process parameter (like temperature, pressure, flow, concentration etc.) that has strong impact on economics or that represents some safety or capacity constraints, for example. The constraints include quality specification, equipment design limits, valve position limits, safety limits, and limits dictated by interlock. The process will be most economical if it is operated very near to its limit or constraints and only infrequently violated. The tolerable frequency of violation is determined by experience, the consequence of violations, and the importance of that parameter and its effect of economics.

As shown in the left-hand side of Figure 5.2, before MPC implementation, the parameter has large variations and the average value of the parameters is kept far away from the constraints as the operator enjoys some comfort margin and also doesn't want to violate the constraints due to large variations. This is represented by a normal or Gaussian distribution at the right-hand side of Figure 5.2. This distribution can statistically be represented by its average value and standard deviations.

$$\text{Distance of average from the constraint} = \beta * \sigma \qquad [5.1]$$

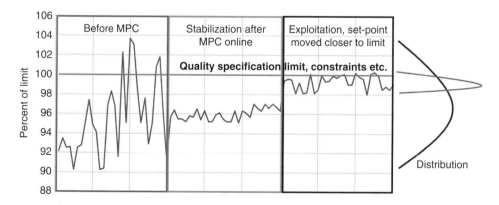

Figure 5.2 Stabilizing effect of MPC and moving of set point closer to limit

Table 5.1 Typical Value of Factor β

Value of β	% of Time Violations Allowed
0.8	20%
1.3	10%
1.65	5%
2.0	2.3%

where

σ is the standard deviation of process variable
β is a factor related to frequency of violation.

From the properties of normal distribution curve, values of β can be found from Table 5.1.

Table 5.1 can be read as $\beta = 0.8$ if violations are allowed 20 percent of the time, and so on.

Stabilizing effect of MPC will allow a shift in the average operating point closer to the limit. It is assumed here that the operating point is moved in such a way that the frequency of violation remains the same as the one that is originally observed before MPC implementation and that is considered as a satisfactory process operation. The calculated economic impact of the shift of the mean close to the operating limit is then used to calculate the benefit due to MPC implementation.

Experience shows that MPC typically reduces the standard deviation between 40 and 70 percent. It depends on process to process, and also depends of effectiveness of implementation team, model accuracy, and so on. if the frequency of violation assume to be same before and after MPC implementation then following mathematical analysis can be done

$$\text{Shift in average operating point} = \beta\left(\sigma_{\text{before}} - \sigma_{\text{after}}\right) = \beta \times \sigma_{before} \times \alpha \qquad [5.2]$$

where σ_{before} and σ_{after} are the standard deviation of process variable before and after MPC implementation. α is the fractional reduction in standard deviation due to MPC.

Because there are a lot of variations in process parameter before MPC implementation, panel operators always try to operate the plant far away from the limit. In other words, operators deliberately do not actively push the constraints, as this will give them some time to take corrective actions during any upsets. On the other hand, MPC pushes the constraints in every minute and MPC equipped with its inherent prediction and stabilization capability. In this way, MPC brings the benefits by reducing the standard deviations of process parameters.

5.3.7.1 Benefit Estimation: When the Constraint Is Known
For cases where the constraint or specification is well known and recorded (like a design pressure or a constant quality specification or an interlock limit), the analysis is

straightforward and simple. Assuming the same frequency of constraint violation, β can be found from the existing plant data (equation 5.1)

$$\beta = \frac{\text{Distance of average from constraint}}{\sigma_{\text{before}}} \qquad [5.3]$$

Substituting this value in equation 5.2

$$\text{Shift in average operating point} = \text{Distance of average from constraint} \times \alpha \quad [5.4]$$

Based on this equation, a certain percentage reduction (β) (typically 0.5) in standard deviation can be applied to the observations of the current plant offset from its known constraints.

5.3.7.2   Benefit Estimation: When the Constraint Is Not Well Known or Changing

For some cases, equation 5.4 cannot be applied directly as the distance of average from the constraint is not known. In those cases, equation 5.2 can be applied directly. For benefit estimation study, experience shows that a typical value of $\alpha = 0.5$ and $\beta = 1.65$ (corresponds to 5 percent violation) can be used. It is believed that benefits calculated by these procedures and assumptions are ±30 percent accurate.

It has to be noted here that this type of benefit estimation procedure should be consider as guidelines, and each and every case has to be considered on their merits and analysis of actual plant data with respect to allowable fraction of constraint violation.

5.3.8   Estimate Change in Key Performance Parameters Such as Yield, Throughput, and Energy Consumption

It is clear that MPC will facilitate shifting the average process value near to its limit or constraints. How this will translate to improve the process performance and how that can be quantified is the main activity of this step. Method of calculation will vary plant to plant, and this has to be discussed with the plant process engineer. A simple example will demonstrate the activity.

5.3.8.1   Example: Ethylene Oxide Reactor

In an ethylene oxide reactor, ethylene and oxygen are reacted in presence of a catalyst to form ethylene oxide (desirable main reaction) and carbon dioxide (side reaction undesirable). Maximum allowable oxygen concentration at reactor inlet is 8 mole percent due to flammability limit. Plant is operating at 7.3 percent average oxygen concentration before MPC implementation. After MPC, the operator is able to shift the average oxygen concentration to 7.8 percent, which is very near to its max limit. Now, oxygen being the limiting reactant, an increase of 0.5 percent oxygen concentration results in 0.2 percent catalyst selectivity increase. This means for a 600 KTA plant, 1ton/hr less ethylene consumption for the same production rate before and after MPC. In this way, a plant will benefit by reducing ethylene flow and running more economically.

5.3.9   Identify How This Effect Translates to Plant Profit Margin

It is absolutely necessary that MPC generates more profit and thus proves itself as a profitable investment. All the process performance improvements should aim to return more profit; otherwise, the MPC project will not be economically viable. The purpose of this step is to quantify all the benefits of yield, throughput, and energy consumption in monetary terms. MPC should not be used only to make the plant more stable with improved control scheme, but at the same time, it should translate those benefits to generate more revenue.

In the above example, 1 ton/hr ethylene savings corresponds to 1 million USD per annum savings. In this way, in this step all the benefits are translated to equivalent money savings. (Good references of industrial applications and benefits calculations can be found in Duarte-Barros 2015; Shin 2006; Sun 2007; Xuesong 2007.)

5.3.10   Estimate the Economic Value of the Effect

The purpose of this step is to make an economic viability study report of the MPC project. Note that an MPC project will be seen as an investment by company top management. Like any other investment, top management will want to know the return of investment before deciding whether to implement MPC. The purpose of this step is to summarize all the potential profitable areas where MPC can be deployed successfully and fetch more economic benefit. All the benefits should be quantified in money terms and added in a payback period or return on investment framework. Company top management wants to see the payback period and a cost–benefit analysis. The purpose of this report or presentation is to present a clear picture of the investment necessary and profit expected that will enable the top management to make the correct decision on MPC project implementation.

5.4   Case Studies

5.4.1   Case Study 1

We want to maximize C4 in LPG in a refinery debutanizer.
    Data are as follows:

Feed 700 KT/year
LPG price 200$/T, C4 price 130 $/T
Current C4 in bottom mean 2 wt. %, standard deviation 1 wt. %
Assume 50 percent reduction in variability is possible by MPC implementation and
    plant operating hours 8000 hrs/year.
Estimate the benefit.

5.4.1.1   Benefit Estimation Procedure
Incentive to increase C4 in LPG = 200 − 130 = 70 $/T
Benefit = Feed × Upgrade × Upgrade value = $600 \times (0.02 - 0.01) \times 70 = 490$ K$/year
Plant availability is 8000 hrs in 8760 hrs.
Actual benefit = $490 \times 8000/8760 = 447.4$ $/year

5.4.2 Case Study 2

The following case study shows how the above procedure can be utilized to calculate the benefit of a simple system. Consider two typical distillation columns, de-ethanizer and de-propanizer normally available in a refinery. Following data are available:
Quality spec:

Ethane wt. % in propane – 3.2% (max)
Butane wt. % in propane – 1.5% (max)
Propane wt. % in butane – 1.5% (max)

The C2s mainly go to the overhead of the de-ethanizer (which end up in fuel gas) but some slip through to the propane.

Flow rate of propane is 75 TPD and Butane 130 TPD.
Value of fuel gas = $75/ton
Value of propane = $200/ton
Value of butane = $180/ton

5.4.2.1 Benefit Estimation Procedure

The first step is to collect 60-minute average data from the data historian of actual plant (if online analyzer is available) or from offline laboratory analysis through LIMS and then calculate the average and standard deviations of these quality parameters.
For sake of demonstration, say the values shown in Table 5.2 are found.
As the propane value is much more than fuel gas, it is beneficial to configure MPC in such a way that it maximizes the ethane content in propane.
The benefit quantification on the de-ethanizer column can be calculated as follows:

$$\text{Shift in average value due to MPC} = \beta\left(\sigma_{before} - \sigma_{after}\right) = \beta \times \sigma_{before} \times \alpha$$

where σ_{before} and σ_{after} are the standard deviations of process variable before and after MPC implementation, and α is the fractional reduction in standard deviation due to MPC.
Assume a 5 percent violation is allowed. Then a typical value of $\alpha = 0.5$ and $\beta = 1.65$ (corresponds to 5 percent violation) can be used.

Table 5.2 Average Value and Standard Deviation of Quality Parameters

Quality Parameter	Average Value	Standard Deviation
Ethane wt.% in propane	2.3	0.4
Butane wt.% in propane	0.5	0.2
Propane wt.% in butane	1.1	0.4

Shift in average value due to MPC $= 1.65 \times 0.4\% \times 0.5 = 0.33\%$

$$\text{MPC benefit} = (\text{Shift in average value}) \times (\text{Flow rate of propane TPD}) \times (\text{Days in a year})$$
$$\times (\text{Propane value} - \text{Fuel gas value in \$/ton})$$
$$= (0.33/100) \times (75 \text{ TPD}) \times (365 \text{ day/year}) \times (200 - 75)\$/\text{ton}$$
$$= 11292 \$/\text{year}$$

The control objective in de-propanizer should maximize C4 in propane, as propane is more expensive than butane.

The shift in C4 in propane is $= 1.65 \times 0.2 \times 0.5 = 0.165\%$

$$\text{MPC benefit} = (0.165/100) \times (75 \text{ TPD}) \times (365 \text{ day/year}) \times (200 - 180)\$/\text{ton}$$
$$= 903 \$/\text{year}$$

References

Duarte-Barros, R. L., & Park, S. W. (2015). Assessment of Model Predictive Control Performance Criteria. *J. Chem*, 9, 127–135.

Shin, J., Byun, K., Joo, E., Yen, M., Yoon, J., Nath, R., & Ahn, S. (2006, October). Model-based Advanced Control Solution Saves Energy on VCM Plant. In 2006 SICE-ICASE International Joint Conference (pp. 930–932). IEEE.

Sun, M., Sun, Y., Feng, N., & Ma, C. (2007, December). Advanced RMPCT control strategy in CDU. In Robotics and Biomimetics, 2007. ROBIO 2007. IEEE International Conference on (pp. 2273–2277). IEEE.

Xuesong, W. (2007). The Application of Advanced Process Control Technology in Refinery [J]. *Automation in Petro-Chemical Industry*, 1, 012.

6

Assessment of Regulatory Base Control Layer in Plants

6.1 Failure Mode of Control Loops and Their Remedies

Building advanced controls on poor basic controls is like building skyscrapers on quicksand. MPC cannot work efficiently if the base control layer or regulatory control layer is weak. It is advisable not to attempt advanced control on top of ill-configured or poorly tuned basic controls. Hence, the strengthening base control layer is an important prerequisite to build the good foundations of MPC.

It is important to understand that basic controls are profit centers, not a cost center. Improving the base control layer immediately increases profitability even before MPC implementation. It is reported that many vendors had achieved 15 to 25 percent of the MPC profit just by improving the base control layer. Reduction of variability in process parameters can be achieved by enhancing the regulatory controller performance. Reduced variability drives efficiency and thus increases profit.

In spite of knowing this, it has been reported that as many as 60 percent of all industrial controllers have performance problems (Bialkowski, 1993; Desborough & Miller, 2002; Ender, 1993). Base layer control can fail in many ways. As it consists of sensors, controllers, control valves, and the process itself, any malfunction of any of them can negatively impact the performance of whole control loop (refer to Figure 6.1). It is important to understand different common failure mode of valves, sensors, and controllers before doing a control performance assessment of base layer. Poor control performance in industrial processes can be caused by one or more of the effects, as shown in Figure 6.1. The following sections discuss potential problems.

6.2 Control Valve Problems

If instruments are the eyes and controllers are the brains, then control valves are the hands of the process industries. Control valves are moving 24/7 to control the process parameters precisely, and everybody wants them to do their work maintenance free. Since control valves are mechanical devices with moving parts, they are bound to fail, and their performance will deteriorate over time. A valve is often the weak link of the entire control loop. It is important to detect valve performance and troubleshoot problems so that optimal control performance can be achieved.

Multivariable Predictive Control: Applications in Industry, First Edition. Sandip Kumar Lahiri.
© 2017 John Wiley & Sons Ltd. Published 2017 by John Wiley & Sons Ltd.

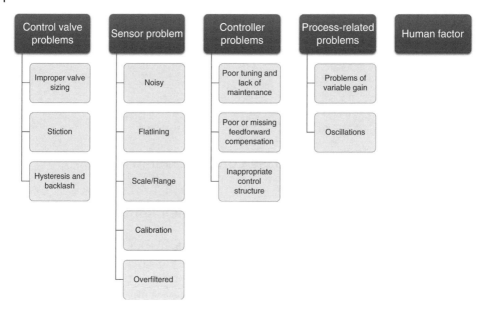

Figure 6.1 Various probable reasons of failure of control loops

There are many ways for control valves to degrade process performance. Most common are improper valve sizing, stiction, hysteresis, and backlash, as shown in Figure 6.1. It has been reported from chemical industry that typically over 30 percent of control valves have problems, many of which can be detected during normal operation using simple tests. Correcting valve problems can significantly improve bottom-line results.

6.2.1 Improper Valve Sizing

Due to economic reasons, most process plants usually operate more than their nameplate capacity. Detailed engineering designers keep this in mind and often build "fat" (excess size) into the design for that contingency. In addition, some plant managers specify, buy, and maintain bigger valves than are needed in anticipation of future needs. These oversized valves result in imprecise control and instability.

Undersized valves are often the result of inaccurate process specifications or when plant is running more than 100 percent of design. Some of the control valves remain fully open when plant is continually running more than 100 percent capacity and thus pose a bottleneck to the process.

6.2.1.1 How to Detect a Particular Control Valve Sizing Problem

Sizing problems can be detected by conducting a few output changes with the controller in manual mode, or by using set point changes in automatic mode. Make at least two process changes in each direction, and the larger the change made, the better (see Figure 6.2). Valve sizing problems exist when the overall process gain is less than 0.3 or greater than 3.

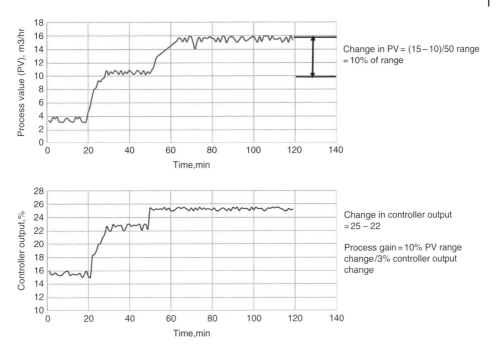

Figure 6.2 Valve sizing problem detection by process gain

Calculate process gain by the following formula:

$$\text{Process gain} = \frac{\dfrac{\left(\text{Final value of PV} - \text{Initial value of PV}\right)}{\text{Transmitter range in engineering unit}}}{\% \text{ change in controller output}}\, \text{in } \%$$

If the process gain number is greater than 3 %PV/%CO or less than 0.3%PV/%CO, then there is a valve sizing problem or a transmitter-scaling problem.

Depending on the cost involved, changing the entire valve or changing the valve trim are the solutions for this case.

6.2.2 Valve Stiction

Valve stiction in a control loop is a very common problem in industry. Identifying stiction as a problem is the first step to combating it.

6.2.2.1 What Is Control Valve Stiction?

Stiction is the resistance to the start of motion of control valve. For control valves having stiction, valve stems will not move when small changes of controller output are attempted and then the stem moves suddenly when there is enough force to free it. Compare this to moving a piece of furniture. You apply increasing pressure and it suddenly starts moving rapidly. The result of stiction is that the force required to get the stem to move is more than is required to go to the desired stem position. In presence of stiction, the movement is jumpy.

The word *stiction* is a combination of the words stick and friction, created to emphasize the difference between static and dynamic friction. Stiction exists when the static

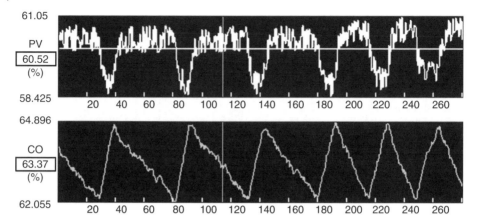

Figure 6.3 Typical trends when valve stiction presents

(starting) friction exceeds the dynamic (moving) friction inside the valve. In a control loop, stiction cause process variables to move in an oscillatory motion, since the controller will push to move the valve until the process variable reaches the set point. If stiction is present, the valve will move too much and the process variable will overshoot. Then, the controller output will reverse direction and it will happen again. Figure 6.3 shows typical trends when valve stiction presents.

6.2.2.2 How to Detect Control Valve Stiction Online
Here are the steps:

1) Put the controller in manual with the output near the normal operating range.
2) Start recording data of controller output (CO) and process value (PV).
3) Change the controller output by 5 to 10 percent to overcome the hysteresis on the loop. If the process variable does not move from this change, keep on increasing the controller output in steps until process variable moves.
4) Allow sufficient time for the process variable to settle.
5) Increase controller output by 0.2 percent in the same direction as the last step. Wait and watch. See if the process variable moves.
6) Repeat the last step until the process variable moves.

The stiction in the loop can be quantified as the total amount of controller output change required to make the process variable move.

6.2.2.3 Combating Stiction
Repairing the valve or positioner to eliminate the stiction is the best way to resolve it. For critical valves, this cannot be done in a running plant as it may require shutdown. In these cases, the better option is to try to minimize the negative effect of stiction.

6.2.2.4 Techniques for Combating Stiction Online
If the valve cannot be repaired, these tuning techniques can be used to keep the plant running:

1) Tune the positioner using a large proportional gain, and no integral action. If derivative action is available, use some to make the valve continuously move. With integral

action in the positioner, the positioner will keep on increasing the controller output when the valve is not moved. After some period of time, the stem will jump, after the positioner has wound up enough. By removing integral action from the positioner, this windup problem is eliminated.

2) If a smart positioner is used, it has the facility to take valve opening feedback, and depending on that, it will adjust the air pressure on the valve.

3) Configure the PID controller (for the control loop) in such a way that the integral action strength can be varied. If absolute error is smaller than some value, then take out the integral action otherwise use it: For error < Threshold value, $K_i = 0$; if not K_i = normal value. Using this method, when the valve is within the stiction band, the integral action is missing from the controller, the controller output will not integrate, having the end effect of removing the stiction cycle from the loop.

4) Use a PID with gap; if the absolute error is smaller than threshold value, the controller output is frozen; if not, the amount of error from the gap is used as the controller input.

6.2.3 Valve Hysteresis and Backlash

When related to a valve, hysteresis is the difference between the valve position on the upstroke and its position on the down stroke at any given input signal. Hysteresis refers to overall response, and backlash refers to that portion of hysteresis caused by lost motion on valve and positioner mechanical parts. By far, the most common causes of hysteresis and backlash are loose or worn mechanical linkages between the positioner, actuator, and/or valve.

Figure 6.4 shows typical trends of the process having hysteresis and backlash. Increasing period as shown in Figure 6.4 is the sign of hysteresis. Effects of hysteresis and backlash are seen in process cycling around the set point, slower controller response, and/or a PV behavior different on "up" movements of the controller output than on its "down" moves.

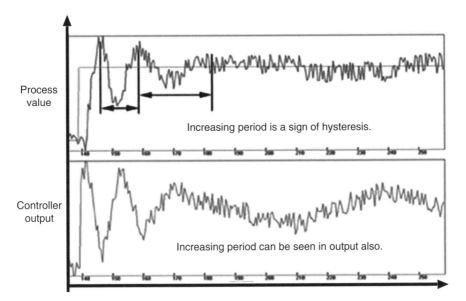

Figure 6.4 Typical trends of the process having hysteresis and backlash

Detecting hysteresis and backlash with the loop in manual (open-loop) mode requires introducing at least two small changes to the controller's output in each direction, and then observing the results. Hysteresis of 1 percent or less is normal in a well-performing control valve; hysteresis greater than 3 percent should be corrected as soon as practical.

Sometimes aggressive controller tuning is the cause of hysteresis in the loop. Due to aggressive tuning, overshoot of process variable and an ever-increasing cycling period is seen (see Figure 6.4).

It is not possible to determine the amount of hysteresis with the loop in automatic mode, though it is possible to observe its existence.

Adding a positioner to a control valve can remove or minimize the impact of hysteresis. If a positioner exists, conduct a thorough physical inspection of sources of lost motion such as positioner, actuator and valve linkages, and repair as needed.

6.3   Sensor Problems

Sensors are the eye of the process industries. Modern chemical plants possess thousands of sensors and generate data each minute. Detecting faulty sensor among those large numbers is itself a challenge. Some of the common sensor failures mode are given as follows.

6.3.1   Noisy

Noisy sensors are very common bad actors of chemical plants. The indications of process parameters revolve around a steady set point. This causes controllers to move the control valve continuously, which subsequently introduces noisy disturbance in the process and affects its smooth operations. Spiking is an indication of another problem.

6.3.2   Flatlining

Some instruments do not respond at all with process changes. They can be detected with straight-line trends in DCS. They are dead instruments.

6.3.3   Scale/Range

Proper scale and range of instruments play an important role to give accurate indications. It often happens that plants continue to run more than 100 percent capacity due to business reasons. Range of instruments were not changed, so some process values are more than their instrument limit or very near to 100 percent high limit. This will impact the accuracy and reliability of sensor measurements. Out of few thousands of instruments, it is very difficult to identify such "under-range" instruments, which subsequently affects the control system performance.

6.3.4   Calibration

Periodic calibration of instruments is necessary to maintain the accuracy of its indications. Instruments that are remotely located (say, pressure transmitters at tall distillation column tops) are often neglected and not calibrated for years. *Drift*—a gradual reduction in instrument accuracy—happens, sometimes unnoticed by operators, which ultimately negatively affects the whole control system performance.

6.3.5 Overfiltered

Sometimes, indications are overfiltered in DCS to show them as a smooth curve. By doing so, control engineer may miss some important process variations, which could have been useful for troubleshooting process problems. It is difficult for panel operator to identify such sensors. Finding such instruments among more than thousands of sensors is a challenge by itself and requires effort and dedication.

6.4 Controller Problems

There are lot of controller problems reported in industry. Some of them are explained below.

6.4.1 Poor Tuning and Lack of Maintenance

Poor tuning of controller is the biggest bad actor in industry for bad controller performance. This may be due to the fact that the controller has never been tuned or that it has been tuned based on a poor model, or even an inappropriate controller type has been used. Typical statistics of control system performance in plants are given in Table 6.1.

It is very common in chemical plants that controllers are tuned during the commissioning stage. After that, tuning is not done for decades, and tuning parameters remain unchanged or kept at default value. However, the performance of many control loops decays over time owing to:

- Changes in the characteristics of the material/product being used
- Modifications of operating points/ranges
- Changes in the status of the plant equipment (wear, plant modifications)

6.4.2 Poor or Missing Feedforward Compensation

If not properly addressed, external disturbances may deteriorate the control performance. If those disturbances can be measured, a feedforward control scheme can be formulated

Table 6.1 Typical Performance of Control Loops in Industry

Sl no	Control Loop Status
1	10% to 90% control loops in manual
2	30% loops-tuning is completely wrong
3	40% loops oscillating
4	Improper PID loop configuration
5	30% DCS systems
6	95% PID on PLCs
7	85% Suboptimal tuning
8	75% Control loops increase variability

to enhance the performance of the control loop. Assessing measures for feedforward control have been proposed by Petersson, Arze´n, and Hagglund (2001, 2002).

6.4.3 Inappropriate Control Structure

It is important to judge whether the control structure is appropriate for a particular control loop. Inadequate input/output pairing, ignoring mutual interactions between the system variables, competing controllers, insufficient degrees of freedom, the presence of strong nonlinearities, and the absence of time-delay compensation in the system are frequently found as sources for control structure problems.

6.5 Process-Related Problems

Sometimes process-related problems can degrade the performance of single-loop controls. In general, these factors fall into two groups: (1) changes in process characteristics (variable gain); and (2) changes in operating conditions.

6.5.1 Problems of Variable Gain

Gain is defined as ratio of an output change to an input change. Each element of a loop—the controller, the actuator, the process, and the transmitter—has a gain. Most of the time, overall gain is constant and linear for a particular operating range. However, if plant capacity changes too much, the response to an input change might vary with the operating point, and this adds a nonlinearity—a variable gain—to the loop.

Possible reasons for variable gain are given below:

- Most of time, gain of the actuator is nonlinear and not constant.
- Valve characteristics, damper curves, and pump curves typically have a nonlinear relationship between percent position and fluid flow.
- In some instances, dead time of process is found changing with flow rate or plant throughput. This causes process gain to vary with plant throughput.
- For level controls, changes in vessel geometry often affect level response.
- Process gains also often change with operating points. Higher temperatures affect reaction rates and yields.

Gain scheduling, nonlinear gain compensation, and self-tuning controllers are some of the solutions of variable gain problems.

6.5.2 Oscillations

Continuous oscillations in process parameters are often seen in the chemical plants. Reasons are many, including damaged actuators, improperly tuned controllers, variable loop gains, interacting loops, and inadequate control strategy designs. If the controller is tuned with very high gain, it can generate expanding oscillation. This can be corrected by reducing the controller gain.

Chemical process industries have highly interactive large number of control loops. Due to these interactions, oscillations in one loop often also appear in other loops as well, through one-way or reciprocal interactions. Similarly, when there are disturbances in uncontrolled variables, due to high interactions, they also appear as variations in whatever PVs they affect. These are the sources of many unexplained oscillations. It is very difficult to identify the root cause of such oscillation—that is, which controller is actually generating the oscillation that ultimately propagates the oscillation throughout many control loops.

Resolving oscillation issues: Variable gain is responsible for oscillation. To pinpoint the oscillation source, investigate which control valve has variable gain in the whole plant or which process can exhibit variable gain.

6.5.2.1 Variable Valve Gain

The relation between valve position and flow is often nonlinear, so valve gain changes with operating point. A nonlinear compensator is often the solution.

6.5.2.2 Variable Process Gain

Process gains often vary with throughput and/or other operating conditions.

Examples include pH loops, and level control in oddly shaped tanks. Programmed or adaptive tuning is often the solution.

Matrices are advised to calculate and help to track down the root cause (see Jelali (2006) and Hagglund (1995)):

- Oscillation period
- Fourier transform
- Oscillation shape
- Comparisons of oscillations of different loops visually

These are discussed in Chapter 15 in great details. Control loop performance monitoring software can help by identifying which loops are oscillating at similar frequencies. This can identify the loop that may be the source of the problem.

Another control loop performance monitoring tool for solving interaction problems is the process interaction map. This tool illustrates the degree of interaction and the relative time-shift of the interaction through a matrix-like diagram. You can pinpoint the root cause as a strong influence (strong color) on the leading side of the diagram.

When you uncover these interactions, modifications to the control system, such as feedforward or decoupling controls, are often the solution.

6.6 Human Factor

All bad controller performance can ultimately be boiled down to the human factor.

It is very common in many chemical industries that control loop performance issues existed for years without anyone addressing them. Operating persons gradually accept the control loops problem as normal phenomena and learned to live with it.

However, if they put effort to resolve those problems, the return is huge. The main aim of control system performance assessment is to evaluate the health of the existing control loops and resolve any issues immediately. Typically, there is no shortage of "low-hanging fruit" that will yield large economic benefits for relatively small investments of time and effort.

6.7 Control Performance Assessment/Monitoring

Over the years, process industries technical community has realized the importance of monitoring the base control loop performance. There are huge benefits in detecting the weakly performing control loop and subsequently improving their performance.

Before attempting to start MPC controller design, a complete detailed assessment of the existing control loop in the plant is necessary. The performance assessment of the existing controller, control valve, and sensors should be broad and may start with getting answers to the following questions:

- How many PID controllers are there in the plant under considerations?
- How many of them are in manual mode or not at their desired mode?
- What are the reasons to keep them in manual mode?
- Is there any effort to put back these controllers in their desired mode in recent past?
- If yes, what are the results and observations? Why were they again brought back to manual mode?
- For controllers that are in their desired mode of operation, what percentage of time are they in their desired mode?

6.7.1 Available Software for Control Performance Monitoring

Due to the large number of control loops presents in any moderate-size process industries, manual evaluation of each control loop performance is not feasible. Online systematic performance monitoring of control loops through various KPIs and matrices is the need of the hour. This gives rise to a new technology/software called control performance monitoring/assessment (CPM/CPA). In short, CPA is an important asset-management technology to maintain highly efficient operation performance of automation systems in production plants.

The main aim of CPA is to provide an online automated procedure that periodically generates reports to plant personnel about the health of the control loops and how each controlled variable is actually performing against its expected performance. The term *monitoring* means tracking the different statistical matrices for changes that reflect the control performance over time. The term *assessment* refers to the action of evaluating control performance at a certain point in time.

Nowadays, a wide variety of commercial performance monitoring tools are available on the software market. Among the most often used are the ProcessDoctor™ from Matrikon, ABB's Loop Optimizer Suite, Honeywell's Loop Scout™, Invensys's Performance Watch, Emerson's EnTech Toolkit, and Control Arts' Control Assessment Tools. All of them use statistics-based indices, autocorrelation function, and various deterministic indices.

Basically these software products define many KPIs or smart metrics, which are online *measurable, meaningful,* and *actionable.* The basic idea is to historize, track, and drill down all these KPIs. The software products build their diagnostic capability based on these metrics. They can automatically detect instrument and valve failure, weakly performed control loops, and causes of oscillations, among other problems.

6.7.2 Basic Assessment Procedure

Jelali 2006 provides an excellent overview of control performance assessment technology along with its industrial applications. A comprehensive approach for CPA should include the following stages:

1) *Evaluate current performance:* Measured (dynamic) data are analyzed and used to calculate various performance KPIs to evaluate the current performance of running control system.
2) *Set the benchmark:* This step actually sets a standard benchmark against which the current control performance will be evaluated. This may be the minimum variance (MV) (as an upper but not achievable performance bound) or any other user-specified KPIs, which can be treated as desired or best possible performance given the existing plant and control equipment.
3) *Detect underperforming loops:* Current control performance is judged against the selected benchmark. Moreover, one can determine the improvement possible by reducing the current performance gap from benchmark. Only those control loops, which are not adequately performing and offer potential benefit, are considered in the subsequent diagnostic steps.
4) *Investigate the underlying causes:* When poor performance is detected, the reasons for this should be investigated by evaluating various KPIs, described in next section and Chapter 15. The diagnostic step is the most difficult task of CPA, requiring experience, expertise, and domain knowledge. Some of the case studies can be found in Thornhill and Hagglund (1997) and Schafer and Cinar (2002).
5) *Suggest improvement measures:* After troubleshooting the problem, corrective and preventive actions must be taken to restore the control loop performance. In most cases, poor working controllers can be improved by retuning and changing valve or transmitters. In some rare cases, desired control performance is not possible with the current process and control structure, then some modifications of control structure can be tried.

6.8 Commonly Used Control System Performance KPIs

There are many metrics employed by different CPA software vendors to quantify the performance of a control system. Also quite a number of performance measures for assessing controller performance have been proposed in the literature, especially during the last decade. Most of them targeted to be computed from normal operating data only.

The performance of a control system relates to its ability to deal with the deviations between controlled variables and their set points (or desired values). These deviations can be quantified by a single number, the performance index (indicator/potential/metric).

Performance KPIs can be grouped by following categories for easy understanding:

1) Traditional KPIs
2) Statistical-based metrics
3) Business/operational metrics
4) Advanced indices

6.8.1 Traditional Indices

6.8.1.1 Peak Overshoot Ratio (POR)

The peak overshoot ratio (POR) is the amount by which the process variable surpasses set point. An aggressive controller can increase the amount of overshoot associated with a set point change.

6.8.1.2 Decay Rate

A large decay rate is associated with an aggressive controller, and visible oscillations are present in the set point response. The smaller the decay rate, the faster the oscillations will be dampened.

Popular values include a 10 percent POR and a 25 percent decay ratio.

6.8.1.3 Peak Time and Rise Time

These measurements gauge the time response to a change in the set point. A large peak and rise time could be the result of a sluggish controller.

6.8.1.4 Settling Time

The settling time is the time for the process variable to enter and then remain within a band. Time spent outside the desired level generally relates to undesirable product. Therefore, a short settling time is sought.

6.8.1.5 Integral of Error Indexes

Other closed-loop performance metrics include the integral of error indexes, which focuses on deviation from set point. The integral squared error (ISE) is very aggressive because squaring the error term provides a greater punishment for large error. The integral time absolute error (ITAE) is the most conservative of the error indexes; the multiplication by time gives greater weighting to error that occurs after a longer passage of time. The integral absolute error (IAE) is moderate in comparison to these two. Additional indexes can be derived, depending on the system requirements. Integral time squared error (ITSE) combines the time weighting with the exaggerated punishment for larger error.

The formulas for calculating the integrated error indexes are listed below:

$$IAE = \int_{0}^{T} |e(t)| \, dt$$

$$ISE = \int_{0}^{T} e^2(t) \, dt$$

$$ITAE = \int_{0}^{T} t |e(t)| \, dt$$

$$ITSE = \int_{0}^{T} t e^2(t) \, dt$$

6.8.2 Simple Statistical Indices

The term *simple indices* refers to indices that can be evaluated with simple statistical calculations and that do not require any nontrivial a priori knowledge. Descriptive statistics are separated into three categories: measures of central tendency, measures of spread, and measures of shape. These statistics are most commonly calculated for the process variable, controller output, and controller error. Various simple statistical indices are given next.

6.8.2.1   Mean of Control Error (%)

The control error mean should, of course, be centered on zero with no offset and a sufficiently small standard deviation.

6.8.2.2   Standard Deviation of Control Error (%)

Long or excessive deviation can easily be identified from the plot of mean of control error (%) and standard deviation of control error (%).

6.8.2.3   Standard Variation of Control Error (%)

$$\text{Standard variation} = \frac{\dfrac{\left[\sum |PV - SP|\right]}{n-1}}{\text{Average } PV}$$

Using this method, a smaller standard variation will represent less deviation from set point. Some factors that can impact the standard variation include the number of set point changes as well as the number of disturbances that impact the process. The standard variation can be used to gauge performance improvement relative to retuning a loop. If the value for standard variation is smaller after retuning, then performance has been improved. It should be noted that when comparing a before-and-after performance index, the data must be collected for a sufficient period of time such that the number of disturbances impacting the system are approximately equal.

6.8.2.4   Standard Deviation of Controller Output (%)

This will give how much variation is in controller output.

6.8.2.5   Skewness of Control Error

Another interesting statistics is the control error skewness, which represents nonsymmetrical data distribution. Skew data often indicate problems of nonlinear character. Sometimes loops exhibit control valve stiction, regularly resulting in the high value of this KPI.

6.8.2.6   Kurtosis of Control Error

For Gaussian signals, the kurtosis should be centered around zero. In this case, if it is clearly around -1 or $+1$ instead, indicating rather non-Gaussian signals. The reason is slow periodic cyclic behavior of the loop.

6.8.2.7   Ratio of Standard of Control Error and Controller Output

$$R = \frac{\sigma_{SP-PV}}{\sigma_{OP}}$$

6.8.2.8 Maximum Bicoherence

Usually, maximum bicoherence of the control error is plotted for all loops. In Choudhury et al., 2004 it is shown that the bicoherence plot can be used to assess signal nonlinearity. Single evaluations may tend to contradict this hypothesis but when considering many data sets, it turns out that such a measure may be able to detect loops that exhibit nonlinearity problems.

6.8.3 Business/Operational Metrics

6.8.3.1 Loop Health

Average value of a group of performance indices; see Ruel (2002).

6.8.3.2 Service Factor

Ratio of time in use of a controller to the total time period considered (also known as time-on control); see Kinney (2003) and Bebar, Krug, and Reinig (2004).

6.8.3.3 Key Performance Indicators

Weighted indices combining different performance metrics; see Ruel (2003) and Schulze (2003).

6.8.3.4 Operational Performance Efficiency Factor

Product of availability, performance, and quality of equipment; see Akesson (2003).

6.8.3.5 Overall Loop Performance Index

Combination of some statistics to an index for isolating problem loops; see Xia and Howell (2003).

6.8.3.6 Controller Output Changes in Manual

Large values indicate a control system that a user cannot trust in automatic. The operator has to provide control.

6.8.3.7 Mode Changes

Similarly, large numbers of transfers indicate controllers that cannot handle upsets and/or set point changes. In both cases, tuning and/or strategy changes may be required.

6.8.3.8 Totalized Valve Reversals and Valve Travel

High and low values indicate which actuators need maintenance and which do not. This insight can shorten turnarounds and avoid unplanned shutdowns.

6.8.3.9 Process Model Parameters

By observing the process response to normal operator actions, control loop performance monitoring can parameterize the process model in terms of dead time, capacity lag times, and steady state gains. This can demonstrate a need for retuning, adaptive tuning, and/or nonlinear characterization.

6.8.4 Advanced Indices

This group of indices involves more complex computations and eventually more prior knowledge.

6.8.4.1 Harris Index

The most prominent of these indices is the Harris index that compares actual loop variability to minimum-variance variability, leading to an index between 0 and 1 where 1 equals minimum-variance performance:

$$\text{Harris index} = \frac{\sigma^2_{minvar}}{\sigma^2_{SP-PV}}$$

and then used for comparison purposes against future values. The Harris Index is useful for assessing the output variance due to stochastic disturbances. It cannot give specific information about set point changes, known disturbance variables, settling time, decay ratio, or stability.

6.8.4.2 Nonlinearity Index

Most of the published KPIs, as stated above, assume the system is linear. Nonlinearity enters into the process through valve stiction, hysteresis, and dead band, and can induce unwanted oscillations. Thus, it is recommended to evaluate how linear (or nonlinear) the closed loop under consideration actually is. An excellent survey by Haber and Keviczky (1999) throws light on various nonlinearity test.

Nonlinearity can be quantified by nonlinearity index (NLI) for a 95 percent confidence level (Choudhury et al., 2004):

$$\text{Nonlinearity index} = \left| b^2_{max} - \left(\overline{b^2} + 2\sigma_{b^2} \right) \right|$$

where σ_{b^2} is the standard deviation and $\overline{b^2}$ the average of the estimated squared bicoherence. Ideally, the NLI should be 0 for a linear process (i.e., the magnitudes of squared bicoherence are assumed to be constant or the surface is flat). This is because if the squared bicoherence is a constant at all frequencies, the variance will be zero and both the maximum and the mean will be same. Therefore, it can be concluded that:

- If NLI = 0, the signal generating process is linear.
- If NLI > 0, the signal generating process is nonlinear.

Since the squared bicoherence is bounded between 0 and 1, the NLI is also bounded between 0 and 1. In practice, an NLI value less than 0.01 is assumed to be zero and, consequently, the process is considered to be linear at a 95 percent confidence level. The larger the NLI, the higher is the extent of nonlinearity.

6.8.4.3 Oscillation-Detection Indices

Oscillations in control loops lead to poor performance, and most of them are caused either by aggressive controller tuning, the presence of nonlinearities (e.g., static friction, dead-zone, hysteresis) or (internal and external) disturbances. According to Desborough and Miller (2002), 32 percent of controllers classified as "poor" or "fair" in an industrial survey by Honeywell show problems/faults in control valves. A number of researchers have suggested methods for detecting oscillating control loops.

Hagglund (1995) suggested a methodology to detect oscillating loops, typically caused by too-high friction in control valves, based on computing the integral of absolute error (IAE) between zero-crossings of the control error e, i.e.,

$$IAE_{\text{oscillation}} = \int_{t_{i-1}}^{t_i} |e(t)| dt$$

The generated oscillation index (OI) h formed using $IAE_{\text{oscillation}}$ can be interpreted in the following way (Forsman & Stattin, 1999):

- Loops having $h > 0.4$ are oscillative, i.e., candidates for closer examination.
- If $h > 0.8$, a very distinct oscillative pattern in the signal is expected.
- White noise has $h \sim 0.1$.

6.8.4.4 Disturbance Detection Indices

A disturbance is defined as anything other than the controller output signal that affects the measured process variable. In an interacting plant environment, each control loop can have many different disturbances that impact performance.

6.8.4.5 Autocorrelation Indices

Autocorrelation is used to determine how data in a time series are related. By comparing current process measurement patterns with those exhibited in the past, the nature of disturbances and how they affect a system can be analyzed.

The equation for calculating the autocorrelation relationship is given by

$$Corr(k) = \frac{\sum_i \left[\left(PV(i) - \overline{SP} \right) \left(PV(i-k) - \overline{SP} \right) \right]}{\sum_i \left[\left(PV(i) - \overline{SP} \right)^2 \right]}$$

where PV = measured process data, $\overline{SP}$ = the set point or the series average if there is an offset, k = time delay in samples, i is the sample number.

Autocorrelation values will always range between -1 and $+1$. If data are random, the values will be approximately zero for all time. Any value that is significantly "nonzero" will indicate that the data are nonrandom. A strong autocorrelation will have an initial value near $+1$ or -1 and the trend will be linear, and this demonstrates a pattern where each measurement dictates the next. A moderate autocorrelation is one in which the plot begins below $+1$ (or above -1) and decreases magnitude toward zero but displays noise. An autocorrelation of closed-loop data can also give an estimate of the response time for an isolated disturbance.

6.9 Tuning for PID Controllers

Controller tuning itself is a vast subject. Periodical tuning of base-level PID controllers is essential for maintaining control loop performance.

In industrial practice, there are four main reasons for lack of tuning and maintenance:

1) The commissioning control engineers and not the production engineer tune the controllers until they are considered "good enough." They do not have time to optimize the control. Most controllers are tuned once they are installed, and then never again.
2) Often, the controllers are conservatively tuned (i.e., for the worst case) to retain stability when operating conditions change in nonlinear systems. This leads to sluggish controller behavior.
3) In most plants, there are only a few people responsible for maintenance of automation systems and there is usually no dedicated knowledgeable engineer to tune the controller.
4) Operators and engineers often do not have the necessary education and understanding of process control to be able to know what can be expected of the control or what the causes of poor performance are. They are not well equipped to tune the controller. Various studies indicate that the half-life of good control loop performance is about six months (Bialkowski, 1993).

6.9.1 Complications with Tuning PID Controllers

In industrial systems, PID controller tuning is complicated by four factors:

1) The system has nonlinearities such as directionally dependent actuator and plant dynamics.
2) Various uncertainties, such as modeling error and external disturbances, are involved in the system.
3) Suboptimal tuning may be necessary to cater to changes in the system over time, such as aging and general wear.
4) Commissioning is easiest without load, but the load is often variable and affects the dynamic performance.

As a result of these difficulties, the PID controllers are rarely tuned optimally and the engineers will need to settle for a compromise performance, given the time available for the exercise. This makes system tuning more subjective, with the potential for different engineers to achieve different, suboptimal performance from the same equipment. Poor tuning can lead to mechanical wear associated with excessive control activity, poor control performance, and even poor-quality products.

6.9.2 Loop Retuning

Loop tuning normally refers to the parameter selection of a fixed-order controller such as a PID. Loop tuning is a classical topic of control theory, and thus will not be discussed here in detail. Lots of standard textbooks (e.g., Seborg, Edgar, and Mellichamp 2004) are available on this subject. In process industries, model-based tuning methods (particularly Dahlin's method, IMC, IMC-based PID-tuning) are highly recommended because they provide considerable insight and usually have one (or no) adjustable parameter(s). However, the type of tuning method varies depending on control objectives of set-point tracking, disturbance rejection, or both. Chien and Fruehauf (1990) and Morari and Zafiriou (1989) describe in detail how IMC tuning provides an excellent trade-off between control system performance and robustness to modeling errors and process changes. It is recommended to tune the controller based on set-point tracking control objective as MPC usually write to the set point of PID controller. When the controller has to be tuned primarily for disturbance rejection, other methods such as those proposed by Chen and Seborg (2002) and Skogestad (2003) may be preferred.

6.9.3 Classical Controller Tuning Algorithms

Over the years, classical control tuning methods have come to dominate the chemical industry. It is important to understand the basics of these fundamental classical controller-tuning methods. The following sections provide some summary of these methods.

6.9.3.1 Controller Tuning Methods
Controller tuning essentially means finding some values of P, I, and D (described below), so that the performance of controller is optimal.

P	Proportional Band	$= 100/K_c$	
I	Integral	$= 1/\text{Reset}$	(min)
D	Derivative	$= \text{Pre-act}$	(min)
T	Oscillation Period		(min)

Oscillation period is the peak-to-peak time with zero integral and derivative time.

6.9.3.2 Ziegler-Nichols Tuning Method
Ziegler-Nichols method is a closed-loop tuning method and is normally applicable to PID controllers with an ideal algorithm. The Ziegler-Nichols method consists of the following three steps:

1) Using proportional control only and with the feedback loop closed, introduce a set point change and vary the proportional gain until the system oscillates continuously. This value of the controller gain is called the ultimate gain-K_u.
2) Measure the oscillation period T, which is the time to complete one cycle of the response frequency.
3) Use the parameters given in Table 6.2 for controller tuning.

6.9.3.3 Dahlin (Lambda) Tuning Method
The Dahlin method is an open-loop tuning method, and it consists of the following steps:

1) Set the control loop to MANUAL mode and give a step in controller output.
2) Note the open loop time constant t (min) that is required for the process variable to reach 63 percent of its final steady-state value. MPC model curves can also be used for calculating t.

Table 6.2 Ziegler-Nichols Tuning Parameters

Control Action	Gain	Reset (repeats/min)	Preact (min)
P Only	$K_u/2$	0	0
PI	$K_u/2.2$	$1.2/T$	0
PID	$K_u/1.7$	$2/T$	$T/8$

3) Calculate the process gain, K_p:

$$K_p = \frac{\Delta PV}{\Delta CO}$$

ΔPV and ΔCO are % changes in process variable (PV) and controller output (CO), respectively.

$$\Delta PV = \frac{\text{Change in } PV \times 100}{PV \text{ Transmitter Range}}, \Delta CO = \frac{\text{Change in } CO \times 100}{100}$$

4) Use the following parameters for controller tuning.

$$\text{Gain, } K_c = \frac{Q(1-L)}{K_p L}; \left(K_c \approx \frac{1}{K_p} \right)$$

$$\text{Reset} = \frac{L}{T(1-L)}, \text{repeats/min; } \left(\text{Reset} \approx \frac{1}{T} \right)$$

$$\text{Preact} = 0$$

Where,

$Q = 1 - \exp(-T/t')$
$t' = $ Desired closed loop time constant, min
$T = $ Period of execution of loop (scan time), min
$L = 1 - \exp(-T/t)$
$t = $ Open loop time constant, min

6.9.4 Manual Controller Tuning Methods in Absence of Any Software

In the absence of automatic software of controller tuning, manual tuning is sometimes the only option available. Manual tuning methods still dominate the industry. For the sake of the readers, some simple but effective manual tuning methods are described below. Note that these methods are not optimal but usually give a quick starting point to tune the controllers.

6.9.4.1 Pre-Tuning

Give a step in set point and observe the response. One easy way to do this is by switching the controller to manual, changing the output slightly, say by 5 to 10 percent, and switching back to auto. Remember to switch off the set point tracking during tuning. If the response is not acceptable, then only touch the gain.

To achieve good PID control manually, the following steps shall be carried out:

Step 1

1) Start with low gain, increase it until oscillation starts at a constant rate.
2) Reduce the gain by 33 percent (multiply by 0.67) from this point of oscillation. In the case of unacceptable oscillations, further reduce the gain by 33 percent, but leave some amount of oscillatory behavior for good control.
3) The initial range of gain values is FIC 1, PIC 2 to 4, TIC 1.5 to 2, LIC 1 to 5.

OR

1) Start with a large proportional band and decrease it until oscillation starts at constant rate.
2) Increase the proportional band by 50 percent (multiply by 1.5) from this point of oscillation. In cases of unacceptable oscillations, further increase the proportional band by 50 percent, but leave some amount of oscillatory behavior for good control.
3) Tune the controller with higher gain for absorbing process disturbance like frequency, and speed.
4) Tune the controller with smaller gain if only set point change is required and the process has fewer disturbances.

Step 2

1) Start with small reset, increase it till oscillation starts at constant rate.
2) Decrease reset by 33 percent (multiply by 0.67) from this point of oscillation.
3) The range of reset values is 1/(Oscillation time × 0.62) to 1/(Oscillation time × 0.67).

OR

1) Start with large integral time, decrease it till oscillation starts at constant rate.
2) Increase integral time by 50 percent (multiply by 1.5) from this point of oscillation.
3) The range of values as guide is Oscillation time × 0.62 to Oscillation time × 0.67.

Note: If dead time exceeds 50 percent of time constant, then a PI setting where $P = 0.3/K_p$ and $I = 0.42 \times$ Process dead time may be used.

Step 3

1) Start with low pre-act, increase it till oscillation starts at constant rate.
2) Reduce pre-act by 25 percent (multiply by 0.75) from this point of oscillation.
3) The range of values is 0.22/Reset to 0.25/Reset (or Oscillation time × 0.16).
4) If there is a high frequency noise, consider adding a filter of 5 to 10 percent of the value of derivative time (pre-act) or 0.8 to 1.6 percent of oscillation time. Filter value of less than 5 percent of derivative time has little use.

Table 6.3 shows some recommended ballpark numbers of PID tuning parameters to start with.

Table 6.3 Recommended PID Tuning Parameters

Process Variable	Gain	Reset (Repeats/min)	Pre-act (min)
Flow	0.2 to 2	20 to 200	0
Liquid P	0.2 to 2	20 to 200	0
Gas P	2 to 100	0.02 to 10	0.02 to 0.1
Liquid level	2 to 100	0.01 to 1	0.01 to 0.05
Temperature	1 to 50	0.02 to 5	0.1 to 20
GC	0.05 to 1	0.008 to 0.1	0.1 to 20

6.9.4.2 Bring in Baseline Parameters

Before attempting tuning, it is recommended to calculate the process's lag time, dead time, and gain. To do this, set the controller on manual. Set its output to somewhere between 10 and 90 percent. Then, wait for the process to reach steady state.

Next, quickly give a step change in controller output. The process variable will begin to change too, after a period of time. This period of time is called the process dead time.

The process lag time is how long it takes for the process variable (PV) to go 63 percent of the way to where it will eventually end up. This would mean that if the temperature increased from 100 °C to 200 °C, the lag time would be the time it took to go from 100 °C to 163 °C.

The process gain, or merely the gain, is found by dividing the total change in the PV divided by the change in the controller output.

6.9.4.3 Some Like It Simple

From the process gain, lag and dead times, we can build a simple tuning table for both PI and PID controllers. Table 6.4 comes from a controller design method called internal model control (IMC). Each cell yields a numerical setting that an operator plugs into a controller.

Some controller mechanisms use proportional band instead of gain. Proportional band is equal to 100 divided by gain.

The values in Table 6.4 are for an ideal type controller. The controller computes controller gain, integral time, and derivative time using the formulas shown. Other tables and computational methods, of which there are many, are needed for other systems. The IMC does produce a nice smooth response, and it provides a starting place for optimizing the control loop.

Table 6.4 IMC Tuning Parameters

Controller Type	Controller Gain (no units)	Integral Time (seconds)	Derivative Time (seconds)
PI control	$\dfrac{\tau}{K(\lambda+\theta)}$	τ	not applicable
PID control	$\dfrac{\tau}{K\left(\lambda+\dfrac{\theta}{2}\right)}$	τ	$\dfrac{\theta}{2}$

where

θ = process dead time (seconds), τ = process lag time (seconds),
K = process gain (dimensionless)

$\lambda = 2\theta$ is used for aggressive but less robust tuning and $\lambda = 2\tau + \theta$ is used for more robust tuning

6.9.4.4 Tuning Cascade Control

1) Keep the master about five times slower than the slave. Since the controller gain is approximately equal to the reciprocal of the process gain, this means that the process gain of the master should be about five times greater than the process gain of the slave.
2) Avoid integral or derivative action in slave, use only gain.
3) Avoid derivative action in the master.
4) Lock slave set points between some known values.
5) Decascade and tune the slave.
6) Cascade and tune the master.

Fortunately, there are commercial software products available that know all the tuning rules, all the different PID formulas, and all the latest tuning procedures. Exactly how each works is beyond the scope of this book and in many cases, proprietary.

References

Akesson, I. N. (2003). Plant loop auditing in practice. In VDI-Berichte 1756, Proceedings of the GMA-Kongress: Automation und Information in Wirtschaft und Gesellschaft, Baden-Baden, Germany (pp. 927–934).

Bebar, M., Krug, Ch., & Reinig, G. (2004). Konzept zur Bewertung der Regelgute von Regelkreisen anhand vorliegender Prozess- und Ereignisdaten aus verfahrenstechnischen Anlagen. *Automatisierungstechnische Praxis*, 46, 73–79.

Bialkowski, W. L. (1993). Dreams vs. reality: A view from both sides of the gap. *Pulp & Paper Canada*, 94, 19–27.

Chen, D., & Seborg, D. E. (2002). PI/PID controller design based on direct synthesis and disturbance rejection. *Industrial & Engineering Chemistry Research*, 41, 4807–4822.

Chien, I-. L., & Fruehauf, P. S. (1990). Consider IMC tuning to improve controller performance. *Chemical Engineering Progress*, 86, 33–41.

Choundhury, M. A. A. S., Shah, S. L., & Thornhill, N. F. (2004). Detection and diagnosis of system nonlinearities using higher order statistics. *Automatica*, 40, 1719–1728.

Desborough, L., & Miller, R. (2002) Increasing customer value of industrial control performance monitoring—Honeywell's experience. In AIChE symposium series, No. 326 (Vol. 98, pp. 153–186).

Ender, D. (1993). Process control performance: Not as good as you think. *Control Engineering*, 40, 180–190.

Forsman, K., & Stattin, A. (1999). A new criterion for detecting oscillations in control loops. In Proceedings of the European control conference, Karlsruhe, Germany.

Haber, R., & Keviczky, L. (1999). Nonlinear system identification— Input–output modelling approach (Vol. 2). Dordrecht: Kluwer Academic Publishers.

Hagglund, T. (1995). A control-loop performance monitor. *Control Engineering Practice*, 3, 1543–1551.

Jelali, M. (2006). An overview of control performance assessment technology and industrial applications. *Control Engineering Practice*, 14(5), 441–466.

Kinney, T. (2003). Performance monitor raises service factor of MPC. In Proceedings of the ISA, Houston, USA.

Morari, M., & Zafiriou, E. (1989). *Robust Process Control*. Englewood Cliffs, NJ: Prentice-Hall.

Petersson, M., Arze´n, K.-E., & Hagglund, T. (2001). Assessing measurements for feedforward control. In Proceedings of the European control conference, Porto, Portugal (pp. 432–437).

Petersson, M., Arze´n, K.-E., & Hagglund, T. (2002). Comparison of two feedforward control structure assessment methods. *International Journal of Adaptive Control and Signal Processing*, 17, 609–624.

Ruel, M. (2002). Learn how to assess and improve control loop performance. In Proceedings of the ISA, Chicago, USA.

Ruel, M. (2003). Key performance index. In Proceedings of the ISA EXPO, Houston, USA.

Schafer, J., & C- inar, A. (2002). Multivariable MPC performance assessment, monitoring and diagnosis. In Proceedings of the IFAC world congress, Barcelona, Spain.

Schulze, K. (2003). Ein modularer Ansatz zum Performance Monitoring von kontinuierlichen Produktionsanlagen. VDI-Berichte 1756, Proceedings of the GMA-Kongress: Automation und Information in Wirtschaft und Gesellschaft, Baden-Baden, Germany (pp. 921–926).

Seborg, D. E., Edgar, F. E., & Mellichamp, D. A. (2004). *Process Dynamics and Control*. New York: John Wiley & Sons.

Skogestad, S. (2003). Simple analytic rules for model reduction and PID controller tuning, *Journal of Process Control*, 13, 291–309.

Thornhill, N. F., & Hagglund, T. (1997). Detection and diagnosis of oscillation in control loops. *Control Engineering Practice*, 5, 1343–1354.

Xia, C., & Howell, J. (2003). Loop status monitoring and fault localization. *Journal of Process Control*, 13, 679–691.

7

Functional Design of MPC Controllers

7.1   What Is Functional Design?

Functional design is the proper planning and design of MPC controller to achieve operational and economic objective of the plant.

Several items need to be define in the functional design stage:

1) Define process control objectives: Why build this MPC controller? Both economics, operation, and control objectives need to be clearly identified at this stage.
2) Identify process constraints, including equipment limitation, process safety limitations, quality spec, process limitation, and sensor and instrument related limitations.
3) Define controller's scope: How much of the process should be included in MPC?
4) Variable selection: How will the objectives be achieved? What will be the MV, CV, and DV list? Which CV will be controlled by which MV, and how?
5) Regulatory control issues: Identify and assess basic regulatory control performance. What changes should be made on basic regulatory control of the plant?
6) Scope of inferential calculations: Identify any opportunity to implement the inferential calculations or soft quality estimators.
7) Optimization opportunity: Where will MPC be utilized to get economic benefit?
8) Define LP or QP objective function: How will the controller optimize the process and capture the greatest benefit?

Functional design stage is most important to extract benefit from the controller. There is no standard procedure to follow to do a functional design. It depends on the expertise and experience of MPC vendor or control engineer, plant operating people, and plant process engineers. Successful functional design amalgamates the domain knowledge of process with MPC knowledge. Rigorous process knowledge, control knowledge, and engineering knowledge and expertise are required.

All other steps of MPC projects except functional design steps are quite well defined and easy to follow. Only functional design steps require arts and science (experience and knowledge) of control engineering and process engineering. It has been found that MPC can be applied to two different plants (or refineries) using the same configuration or licensors, and still get achieve different benefits. This is due to the different approaches to MPC functional design and different strategies regarding how MPC exploits the benefit.

Multivariable Predictive Control: Applications in Industry, First Edition. Sandip Kumar Lahiri.
© 2017 John Wiley & Sons Ltd. Published 2017 by John Wiley & Sons Ltd.

Functional design is also sometimes very plant specific and depends on plant base control, equipment limitations, and management objectives. Thus, it is absolutely necessary to discuss the following steps of functional design with panel operators, plant production and process engineers, plant control engineers, planners, and plant management. Through this discussion, many aspects of controller design will be known and formulated. A good description of concepts of RMPCT MPC technology of Honeywell are given in Bowen (2008); Lu (1997); and MacArthur (1996). Other good sources of information are DMC Corp. (1994) and Honeywell Inc. (1995).

7.2 Steps in Functional Design

The steps in functional design are shown in Figure 7.1.

7.2.1 Step 1: Define Process Control Objectives

Every MPC scheme has at least three objectives: economics objectives, operating objectives, and control objectives.

7.2.1.1 Economic Objectives

The main driving force to make an MPC application is to make more profit. If MPC don't make profit, then it is not worth building. Thus, a successful MPC application needs thorough understanding of the plant operation, process constraints, and economics of the process. There are no shortcuts. People who jumped to do a functional

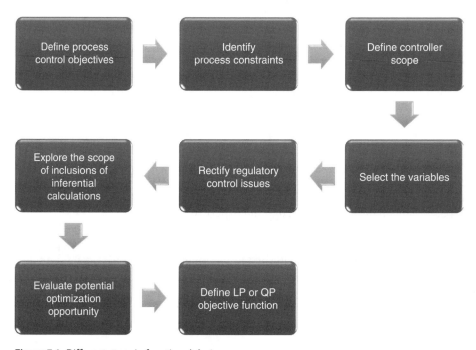

Figure 7.1 Different steps in functional design

design without understanding this end up building an MPC application that does not extract any economic benefit.

MPC application starts with the following considerations:

- Understanding how the unit makes profit is essential to design good controls applications.
- Understanding the process limits and constraints that limit profit maximization is essential to building a successful MPC application.

There is no shortcut and universally acceptable method to address these considerations. However, the following guidelines give an idea of what to look for:

- How does the unit or the plant make profit?
- What are the main products and feed?
- What are the utilities and chemicals used in the process?
- What is the profit margin, and how it is calculated?
- Which product or raw material or utility cost affects the profit margin most?
- How can profit be increased? (Maximize feed, minimize utilities, minimize product quality giveaway, control product qualities to target etc.)

Sweating the asset is the buzzword of modern chemical process industry. Increasing dollars per hour is the main mantra to maximize profit. Normally, maximizing product flow or minimizing utility and raw material brings profit.

What can be implemented to increase profit? Options include (but are not limited to) the following:

- Increase plant capacity within safety boundary limit.
- Maximize most valuable product flow.
- Changing products split: A plant has more than one product; which product flow should be increased to gain maximum profit?
- Reduce utility consumption, and determine how.
- Increase catalyst selectivity or yield.
- Increase severity of operation (how?).
- Increase energy efficiency.
- Reduce fuel gas and/or chemical consumption.
- Increase process efficiency.

7.2.1.2   Operating Objectives

This is what plant operators want to achieve in their day-to-day operation. Again, these are plant-specific:

- Achieve and exceed daily production target as per company business plan.
- Minimize daily throughput variation.
- Reduce down time of the plant.
- Increase catalyst selectivity.
- Increase reactor or furnace severity.
- Minimize waste generation.
- Maximize process-to-process heat transfer.
- Minimize quality give away against specification.
- Minimize fuel consumption in furnace.

- Reduce time off-spec during crude change.
- Make less slops.
- Have more stable product quality for blending/downstream unit.
- Lower pressure operation.
- Use a low ratio of fuel gas to oxygen.
- Have more severe flashing.
- Minimize steam consumption.

7.2.1.3 Control Objectives

To know the control objectives, answer the following queries:

- What is the expectation of plant personnel from control system of the plant?
- Which control loop consumes the maximum amount of the console operator's time?

Again, this is plant-specific, but here are some examples of control objectives:

- Stabilize complex dynamic control loop.
- Minimize fluctuation in pressure and temperature in particular distillation column.
- Reduce variation in product quality.
- Operate all the PID controllers in their desired mode of operation.

7.2.2 Step 2: Identify Process Constraints

Every process has some limitations to increase profit continuously. The purpose of this step is to identify all the constraints associated with the plant, which can limit its profit maximization. Normally, MPC increases plant throughput until hindered by some constraints or equipment limitations. It is well known that MPC fetch the benefit by pushing the process to its limit. Thus, it is absolutely necessary to have a clear understanding of what are those limitations before designing detail controller.

7.2.2.1 Process Limitations

Process limitations include limitation in reaction system, distillation columns, and heat exchangers. Examples include:

- Jet flooding or down comer flooding, entrainment, weeping in distillation column
- Foaming in absorber or stripper, which causes poor separation
- Overloading in process equipment, such as design capacity limitation in compressor
- Pinch and fouling in heat exchanger
- Overhead cooling restriction in condenser during summer

7.2.2.2 Safety Limitations

These include various safety boundaries around the process and normally safeguard by various trips and interlocks. MPC must run the whole process within this interlock boundary:

- Maximum allowable oxygen concentration in a hydrocarbon mixture
- Auto ignition temperature, lower and higher explosive limit (LEL and HEL), of flammable mixture

Equipment limitations: It includes limitations in various process equipment like reactor, distillation column, various rotating equipment, heat exchanger, and so on. Examples are:

- Surging of compressor
- Maximum velocity through heat exchanger tube
- Maximum allowable tube metal temperature (TMT) in furnace
- Rotating equipment loadings (amp)
- Valve opening (pressure drop)
- Heat exchanger performance, etc.

7.2.2.3 Process Instrument Limitations

These include limitations in control valves, limitations in base control configurations, and various process sensors limitations. Examples are:

- Sticky control valves
- Control valves with hysteresis
- Wrong tuning parameters in base control and thus operated in manual mode
- Noisy sensor
- Sensor with large dead time
- Inaccurate and nonresponsive sensors or transmitter
- Faulty analyzers
- Large cycle time of analyzers

7.2.2.4 Raw Material and Utility Supply Limitation

Many processes cannot increase throughput due to raw material or utility supply limitations (say, fuel gas or steam limitations). Nowadays, petrochemical plants are situated in a complex, and it is the top management who allocate raw material or fuel gas or utility among various plants, depending on overall profit margin and market demand. Sometimes it is profitable to run a particular plant at its lowest capacity and allocate that raw material (say ethylene) to another, more profitable plant. This poses serious drawbacks for MPC performance, as one of the major objectives of MPC is to increase profit by increasing throughput. In such situations, MPC scheme is intended to improve efficiency, selectivity, and/or yield, instead of capacity. It is absolutely necessary to know this beforehand and design the MPC scheme accordingly.

7.2.2.5 Product Limitations

Many plants cannot increase products due to a marketing problem. So the business plan was made at the start of the year, depending on market demand and sales price. In such situations, there is no point in increasing plant capacity and produce more than what can be sold.

7.2.3 Step 3: Define Controller Scope

Controller scope means to decide how much of the process will be included in MPC. Guidelines should include enough of the process to cover the significant constraints on accomplishing objectives. Normally, it is advisable to include all of the process. The main deciding factor is economics. Guidelines are as follows:

- If there is scope to improve economics by controlling a process unit, then it should be included.
- If a particular process unit affects any economic objectives of plant or any constraints, then it should be included.
- Include enough of the process to cover the significant constraints on accomplishing objectives.
- Sometimes following may be excluded from the MPC scope.
- End product storage vessels, as for example, ethylene oxide bullet or mono ethylene glycol storage vessels.
- PLC operated automatic processing, as, for example, pressure swing adsorption (PSA) unit, de-ionization (DI) unit for water treatment.
- Where process automation is so extensive (say PLS-based operation) and process conditions are so tight that little variations can be done. This type of section can be excluded from MPC.

Controller scope should also cover whether MPC includes one big controller covering multiple unit or small separate controllers for each unit. The deciding factor is often the interaction between unit and economic gains. The main considerations here are as follows:

- How much downstream units must know about the performance of upstream units, control-wise and optimization-wise.
- Will the knowledge of the upstream unit's performance help greatly to optimize downstream units? If yes, then go for one big controller.
- What is the opportunity cost economy-wise and control-wise if a separate controller is chosen? If there is substantial economic loss, then go for one big controller.

On the other hand, small controllers are easy to manage and design.. The whole process units often break down by operating unit similar to what console or DCS operator considers. Console operators love to operate smaller controllers. If there is no interaction, then create several smaller controllers.

All parts of the controller must be connected. The main issue is to find a way for separate to controllers talk with each other—for example, through feedflow as DV in each controller or by any other means.

But big controllers will optimize the whole process, and will get feedback from all the process events. Big controllers will need extensive computational power and faster computers to execute in real-time basis (typically one minute).

7.2.4   Step 4: Select the Variables

In this step, a preliminary list of manipulated variable, controlled variables, and disturbance variables are prepared. The first step of variable selection is to identify all the process parameters.

7.2.4.1   Economics of the Unit

Variables that affect process economics include product flow, feed flow, different utility or chemical flow, selectivity, yield, and reactor severity. Identify parameters that need to be moved (either maximized or minimized) to increase economic benefit.

If the measurement of some variables is not available (say, reactor severity, yield, selectivity), then proper calculation and inferential measurements should be taken.

7.2.4.2 Constraints of the Unit

Identify and include process parameters that represent process constraints or equipment limitations, as discussed above. Confirm that all the process constraints are suitably addressed.

Examples of variables that represent process constraints include delta P of distillation column, heat exchanger approach temperature, TMT for furnace, amperage for rotating equipment, vibration for compressor, interstage temperature for multistage reciprocation compressor, maximum allowable oxygen concentration in mixture gas, and valve opening.

7.2.4.3 Control of the Unit

Identify and include all process parameters that need to control the process. Also include all the parameters that directly or indirectly represent disturbances entered in the system. Examples are cooling water temperature, atmospheric temperature, temperature, pressure, and level of various equipment in the process.

7.2.4.4 Manipulated Variables (MVs)

Manipulated variables are the control handles on the process. These are the independent variables whose conditions are manipulated (changed) to control the CVs. MPC adjusts MV values to achieve control and optimization objectives. The controller never moves an MV outside its limit.

Practical guidelines for choosing MVs

1) Some MVs are obvious: operators always move them, and they had a control valve associated with them.
2) Typically, manipulated variables (MV) in MPC are the set point of a regulatory controller, like a reflux flow controller set point.
3) If the process is very slow, then use the valve position (instead of controller set point) as the manipulated variable, like the reflux flow control valve opening.
4) Some MVs are not so obvious: There are instances in which operators rarely moves MVs. This does not mean they should not be included (e.g., pump around flows).
5) Be careful about discontinuities such as split range valves and overrides.

Examples of MVs

Reflux flow, reboiler steam flow or duty, overhead pressure, feed temperature, feed flow, compressor speed, and heater fuel gas pressure

7.2.4.5 Controlled Variables (CVs)

Controlled variables are the process conditions to be controlled (that is why they are also called PVs, process variables). The bottom temperature of a distillation column can be CV.

Practical guidelines for choosing CVs

1) Include all the process parameters as CV, which belongs to following three classes:
   Parameters that should be control to improve unit stability
   Parameters that represent operational limit
   Parameters that are used to evaluate economic objectives
2) Ensure that each CV is dependent on at least one MV.
3) CV can be measured CVs or inferred CVs. For measured CVs, consider measurement noise, dead time, and dynamics. For inferred CVs, use data driven or first principle engineering model.
4) Make sure all process constraints are included in the controller's CV list.
5) Include valve opening as CV rather than relying on DCS anti-windup processing.
6) Be careful about parallel CVs.

Examples of CVs

Temperature, pressure, delta pressure, inferential calculations, concentration measured by analyzers, valve position

7.2.4.6 Disturbance Variables (DVs)

Disturbance variables are measured disturbances (changes) in the process that influence the CVs, but that are not under MPC control. DVs often cannot be moved independently. Cooling water temperature and ambient temperature can be DV.

Practical guidelines for choosing DVs

- Any disturbance that has significant effects on CVs but cannot be controlled should be included in DV list.
- Ensure that DVs must be independent of MVs and CVs.
- DVs should have reliable measurements.
- DVs often cannot be moved independently.

Examples of DVs

Feedflow, feed temperature, feed composition, ambient temperature, cooling water temperature

7.2.4.7 Practical Guidelines for Variable Selections

Initially prepare a preliminary list of MVs, CVs, and DVs. Then check the following and discuss with plant operators:

- Ensure that all parameters that affect operating costs of the unit are included in the CV list.
- Ensure that all the constraints are properly identified and directly or indirectly included in the CV list.
- Ensure that all measurable disturbances of the process are included in either the CV or DV list.
- Ensure that all the control valves and their associated controllers are in some way included in the MV list.
- Ensure that all the handles that operators normally use to control the process are included in the MV list.

7.2.5 Step 5: Rectify Regulatory Control Issues

MPC is a supervisory controller that acts above the basic PID regulatory controller. It is just like building a multistoried building on underground foundation. If the foundation is weak, a strong building cannot be made upon it. So it is absolutely necessary to ensure that a basic PID regulatory controller works well before trying to build MPC over it. So a complete assessment of basic level regulatory control is necessary. This is discussed in detail in the previous chapter. This step is again included in the functional design stage to reevaluate and revisit the regulatory control issues if any issues remain unaddressed. This step consists of (but is not limited to) the following items:

1) Assessment of basic control loop, their capability for disturbance rejection, dead time, overshoot, time constant, and so on.
2) Proper tuning of regulatory controller. Poor regulatory controller tuning leads to poor performance of MPC.
3) Sensor/transmitters health checkup for accuracy, repeatability, and noise,
4) Control valves heath checkup for stickiness and hysteresis, as discussed previously.
5) Analyzer health checkup for cycle time and accuracy.
6) Change of regulatory controller strategy to take advantage of MPC.

7.2.5.1 Practical Guidelines for Changing Regulatory Controller Strategy

1) Open cascades that operate with sluggish or cyclic behavior. Opening cascades is preferred, as it gives MPC an extra degree of freedom. By default, cascade should be break or made open unless warranted by maintaining product quality.
2) Examine overhead temperature/reflux cascades and bottom temperature/reboiler heat cascades on a case-by-case basis. Criteria to select the case are as follows:
   Does the loop reject disturbances?
   Does the loop work well (evaluate ITSE, ITAE)?
   Does the loop respond fast relative to MPC?
3) Close simple cascades to remove very fast disturbances (e.g., heater outlet TC to fuel gas FC).
4) Remove any advanced controls in basic regulatory control level that operate in the same time domain as the MPC.
5) Remove control systems that create nonlinear or discontinuous behavior (like split range controllers).
6) Add feedforward measurements where possible to allow better disturbance rejection.
7) Break regulatory level control loop (by default, break it). A well-designed MPC model uses the level inventory as a buffer to stabilize the process. Normally, level control valves adjust the feed to another unit. Thus, to maintain the level at its set point, fluctuation is created in the feed of downstream units. There is no incentive to maintain level at rock steady value; instead, most of the cases level within a range (say, between 30 and 60 percent), which is sufficient for the operation. So it is advisable to break the level and flow cascade loop at column bottom or at reflux drum. It is better to hand over this job to MPC, which will take care of the level of the column and feed flow of downstream units simultaneously.

7.2.6 Step 6: Explore the Scope of Inclusions of Inferential Calculations

It is necessary at this stage of functional design to evaluate the scope to introduce soft sensors or inferential calculations. By definition, an inferential or soft sensor is a mathematical relation that calculates or predicts a controlled property using other available process data. Examples of soft sensors are many:

- Polymer process: Melt flow index (MFI) or melt/flow ratio (MFR), density
- Refining or petrochemical process: Product impurity ($C5^+$, $C4^-$ etc.), petroleum fraction boiling points IBP, EP
- Petroleum fraction cold properties (such as flash, freeze, cloud point), severity, reformer octane etc.

When to consider introducing soft sensors

- If analyzer is not available to measure an important CV (which may have impact on unit economics or represent an important process constraints), it is unreliable, or dead time is prohibited, consider use of soft sensor or inferential calculations.
- If a CV's real-time information and prediction is very important from control or optimization purpose and that is not available, then consider introducing a soft sensor.
- Maybe these CVs are determined as low sampling rate (online analyzer) or through offline laboratory analyses only. In most cases, these variables directly or indirectly affect the process output quality. Thus, the real-time information with high sampling rate is very important for these variables for process control and management.
- For these reasons, it is of great interest to deliver additional information about these variables at higher sampling rate and/or lower financial burden, which is exactly the role of soft sensors.
- Sometimes some CV such as flooding percentage, down comer or tray loading, loading in packed column, or severity in reaction cannot be measured directly but represents important process constraints. In this situation, an inferential calculation or soft sensor will help to include that CV in MPC model.
- When it is very difficult or costly to measure an important parameter online, like distillation tower top product impurity, soft sensors are used to predict that inferential property from other easily measurable parameters such as top temperature or pressure.
- Sometimes, soft sensors are used as backup to an existing analyzer to reduce or eliminate dead time, both from the process and the analyzer cycle.

In this step, critically evaluate the inclusion of soft sensors, which can enhance the MPC controller performance and increase profit.

7.2.7 Step 7: Evaluate Potential Optimization Opportunity

It is important to identify all the optimization opportunities at this stage of functional design. Where the opportunities lie and how to exploit them for economic benefit is to be planned in this step. This step consists of (but is not limited to) the following items:

7.2.7.1 Practical Guidelines for Finding out Optimization Opportunities

- MPC has a stabilization effect. It can reduce standard deviation of key process parameters by 50 percent. If a process is losing some economic benefit due to its instability or fluctuation, then MPC can definitely help to gain benefits in such situation. Find out which area of the process can utilize the MPC stabilizing effect benefit.
- If there are large interactions between process variables due to highly integrated plants, MPC can bring benefits in such plants by utilizing its feedforward predicting capability.
- Large delay processes also benefit from MPC.
- If ambient conditions can impact economic benefit, MPC can exploit that when ambient conditions are favorable.
- MPCs always try to increase plant load, and this is the biggest area of gain. It is not uncommon for a 3 to 5 percent load increase in successful MPC application, as mentioned in the literature.
- By introduction of important soft sensors (like flooding percentage, reaction severity etc.), MPC can push the process to its limit. In this way, MPC can fetch benefits.
- MPC can target multiple constraints and thus extract maximum economic benefit.
- MPC also can change the product distribution as per market demand and profit margin.

7.2.8 Step 8: Define LP or QP Objective Function

MPC has two major objectives. First, it tries to control the process. That is, it tries to maintain all the CVs within their range (or at their set point). Second, if it is able to do that, then it tries to optimize the process. That is, it tries to increase economic benefit while obeying all the constraints.

In many applications, the control requirement to keep all the variables inside constraints does not use up all the degrees of freedom available to the controller. Even when there are more CVs than MVs, there can be extra degrees of freedom if a number of the CVs have ranges rather than set points, which is often the case.

One can put the extra degrees of freedom to good use by defining an objective function so that the controller optimizes some aspects of the process in addition to controlling it. The objective function is a linear or quadratic function of any or all of the CVs and MVs. In this step, the form and the variables of objective function are decided. It will help to formulate a strategy of profit maximization through the use of the MPC controller.

The general form of the objective function is

$$\text{Minimize } Obj = \sum_i b_i CV_i + \sum_i a_i^2 \left(CV_i - CV_{oi} \right)^2 + \sum_j b_j MV_j + \sum_j a_j^2 \left(MV_j - MV_{oj} \right)^2$$

where b_i = linear coefficients on the CVs
b_j = linear coefficients on the MVs
a_i = quadratic coefficients on the CVs
a_j = quadratic coefficients on the MVs
CV_{oi} = desired resting values of the CVs
MV_{oj} = desired resting values of the MVs

The controller minimizes the objective function (or maximizes the negative of it) subject to keeping all CVs within limits or at set points and all MVs within limit.

In many applications, the major product and feed streams and the utility usage are included in the set of MVs and CVs. For such applications, a typical use of the objective function is to maximize operating profit, defined as the value of products minus the costs of feeds and utilities.

In such cases, objective function is defined as

$$
\begin{aligned}
Obj = &+\sum \text{Product flows} \times \text{Product values} \\
&- \sum \text{Feed flows} \times \text{Feed values} \\
&- \sum \text{Utility flows} \times \text{Utility values}
\end{aligned}
$$

7.2.8.1 CDU Example

$$
\begin{aligned}
Obj = &\text{Naptha} \times \text{Naptha unit price} + \text{Kero} \times \text{Kero unit price} + \text{LGO} \times \text{LGO unit price} \\
&+ \text{HGO} \times \text{HGO unit price} + \text{Residue} \times \text{Residue unit price} - \text{Feed} \times \text{Feed unit price} \\
&- \text{Fuel} \times \text{Fuel unit price}
\end{aligned}
$$

This is important to decide the type (LP or QP) of objective function and its terms—that is, which product, feed, and utilities will be included in the objective function. A preliminary objective function is selected in this step of functional design.

References

Bowen, G. X. D. F. X. (2008). Application of Advanced Process Control, RMPCT of Honeywell, in DCU. *Process Automation Instrumentation*, 4, 013.

DMC Corp. [DMC] (1994). Technology overview. Product literature from DMC Corp., July 1994.

Honeywell Inc. (1995). RMPCT concepts reference. Product literature from Honeywell, Inc., October 1995.

Lu, J., & Escarcega, J. (1997). RMPCT: Robust MPC Technology Simplifies APC. In AIChE Spring Meeting, Houston, TX.

MacArthur, J. W. (1996). RMPCT: A new robust approach to multivariable predictive control for the process industries. In The 1996 Control Systems Conference (pp. 53–60).

8

Preliminary Process Test and Step Test

8.1 Pre-Stepping, or Preliminary Process Test

8.1.1 What Is Pre-Stepping?

Pre-stepping is a process to move MVs and DVs before the main step test to observe a change (above the noise level) in the CVs. Step size of the process is such that a change in the relevant CVs is clearly observable: and $\Delta CV \gg CV$ *noise.*

8.1.2 Objective of Pre-Stepping

1) The basic objective of pre-stepping is to save time in the main step test. There may be many obstacles/disturbances to doing a step test. Obstacles include data collection system not performing well, step sizes being too small to make a substantial change in relevant CVs above its noise level, or step sizes being too big, which leads the process to undesired zone, where some of the basic P&ID loop is not tuned well. The idea of pre-stepping is to identify these unforeseen situation in the early stage and rectify them beforehand. This will save time during the step test.
2) Key objective of pre-stepping is to identify the size of moves required in the MVs and DVs and to observe a change (above the noise level) in the CVs.
3) Another objective is to determine the settling time of the system.

8.1.3 Prerequisites of Pre-Stepping

The following things must be checked and made ready before starting the pre-stepping:

1) Ensure that the regulatory controllers are tuned properly. If not, retune them.
2) Ensure all regulatory instruments are running accurately and reliably. This includes online analyzer, pressure, temperature, and flow transmitters. Control valves are sticky free and hysteresis free. Ensure that there is enough margin on transmitter range and control valve opening so that they will not go out of the range or be fully open during step testing.
3) Ensure the functional design and MV, CV, and DV list has been finalized and discussed with operation department before starting pre-stepping.
4) Ensure online inferential calculations are running OK and make sense from underlying physics. Check that inferential calculations track or follow online analyzer and/or laboratory analysis.

Multivariable Predictive Control: Applications in Industry, First Edition. Sandip Kumar Lahiri.
© 2017 John Wiley & Sons Ltd. Published 2017 by John Wiley & Sons Ltd.

5) Ensure the regulatory control loop is as per functional design. Break/close cascade loop, level control loop, and ratio loop as per functional design.
6) Ensure the data collection system is running properly. It is good practice to collect data for all the available tags. Data are free, step test are costly. Ensure that there is enough storage capacity available to save big amounts of data during actual step tests. Collect data as much as feasible.
7) Determine the size of the step for each MV and DV so that a clear CV response above noise level can be observed. Prior experience, a little bit of trial and error, study of historical data, and discussions with DCS panel operators can help to determine the step size. Too low a step size fails to change the relevant CV above noise level. Too big a step size leads the process to disturbance zone, and it is difficult to hold and maintain other MVs during this time, which is a requirement of an open-loop step test.

8.1.4 Pre-Stepping

The activities usually done during pre-stepping stage consist of (but are not limited to) the following:

1) Start data collection (one settling time in advance).
2) Give a predetermined step in MV.
3) Set up short-term trends to historize key information.
4) Check that change in relevant CV is observable and $\Delta CV \gg CV$ noise. Ensure moves are large enough to produce a 6:1 ratio of signal to noise in the key dependents. CVs must lift out of the noise band. If it is within noise band, no identification routine can generate model curves. In such cases, curves will be random and erratic.
5) Step all MVs and DVs.
6) May be one to two steps (one up and one down) for each MV.
7) Wait for the process to settle between moves. Hold all other MVs constant at their value.
8) Determine the settling time of the system from the trend.
9) Identify potential problems, if any.
10) Observe any disturbances or unforeseen situation during pre-stepping. This includes but is not limited to:
    - Nonoptimum tuning of basic controller
    - Instruments or analyzers not running properly
    - Stiction in control valve
    - Valves open or close fully
    - Override control, integrators, safety interlocks actuated during pre-stepping.
    - Some CVs goes beyond their acceptable limit, which leads to foaming/flooding/off-spec product/safety interlock actuation etc.
11) Correct all of these disturbances during the trial of pre-stepping so that the main step test can be performed flawlessly.
12) Develop expectation matrix as shown in Figure 8.1. This is a pictorial matrix form depicting where one can expect a model curve (with positive or negative gain), where it is doubtful, and where not expected at all. Ensure that the preliminary model from pre-step data is directionally correct. Discuss the entry of expectation matrix with operation department before pre-stepping. Ensure that a clear MV-CV model curve can be generated from these data.

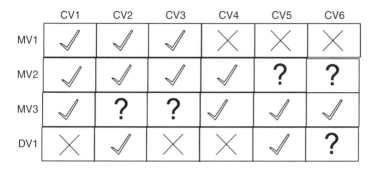

Figure 8.1 Expectation matrix (√ definite response expected, X no response expected, ? response is doubtful)

13) Ensure that the idea of settling time is justified from the curve.
14) Identify preliminary dynamic model. Ensure that model curves of the expectation matrix are as per expectation and there are no surprises.
15) Based on the experience and results of pre-stepping, develop the main step test pattern.

Note that pre-stepping is the last chance to identify and correct all unforeseen controllability problem before the main step test. Ensure that all potential problems are addressed in this step to avoid time loss later on.

8.2 Step Testing

8.2.1 What Is a Step Test?

A step test is the activity where a particular MV is given a step input (either up or down) and the dynamic effect of that step on different CVs are recorded over time (see Figure 8.2). Its purpose is to build the model later on for that particular MV-CV pair. The process is repeated for all MVs and DVs so that all the dynamic effects of an MIMO system can be captured.

8.2.2 What Is the Purpose of a Step Test?

Step test is the most important activity to collect data for model building. As MPC models are data-driven models, a good step test is a prime prerequisite to make a good model, which ultimately leads to effective MPC implementation. The purpose of step testing is to generate data for model building by moving all MVs one at a time and allowing the CVs to drift as an effect of MV change. Each MV should move 15 to 20 times up and down from the normal operating point, and the process is allowed to settle down. The key idea is not to control the CVs at their desired value but to allow them to change over time and capture the change in the model.

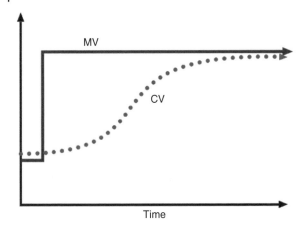

Figure 8.2 Basic concept of step test

8.2.3 Details of Step Testing

There are two aspects of step testing: administrative and technical aspects.

8.2.3.1 Administrative Aspects

Step testing requires considerable time and effort. Sometimes process get disturbed, units run sub optimally, or chances of product get off spec during step testing. So it is necessary to keep plant management informed about the plant step testing and possible outcomes.

Make sure that the plant operation head, planning department, and process engineering department really understand that plants may run suboptimally during step test, which can affect daily production rate, on-spec production, and so on. It is a good practice to review the implications of step testing with all concerned departments and get their green light beforehand. All the potential problem area or impacts should be discussed beforehand.

It is absolutely necessary to discuss in detail the step-testing plan with the operation department. Size and duration of moves should be agreed beforehand with operation and process engineering departments.

A detail meeting with all console operators is necessary before starting the step test. Normally, DCS operators have a tendency to take corrective actions when they see a CV is drifting out from its desired value. It is necessary to explain to the operators that is not the purpose of step testing, so they can allow the CV to drift outside its desired values. Good practice is to explain to the DCS operators the whole idea of step test and dos and don'ts of step testing.

Normally, console operators have a sense how much they can allow the CV to drift before it can affect product quality and plant operation. It is necessary to develop a communication channel with console operator so that they can give a warning signal in real time to MPC engineer in charge if such cases happen during step testing.

8.2.3.2 Technical Aspects

- Ensure that all the concerns identified in the pre-stepping test are addressed properly.
- Provide engineering coverage for the whole duration of the step test. It is good practice to monitor the process parameters from DCS console 24 hours/day.

- Provide a step test plan to DCS operators that describes which MV to move when, how much, and how long to hold other parameters. Also provide approximate settling times and which CVs will be allowed to drift.
- Provide a step test log and a tracking sheet. In the tracking sheet, monitor how well progress is being made in relation to the plan. In the step test log, record times of upsets and shift changes and every detail of the operation during step changes that can affect the CV-MV relationship.

8.2.4 Different Step-Testing Method

8.2.4.1 Manual Step Testing

This is most common method, and it is widely used. In this method, usually one MV moves at a time as per plan, while keeping all other MVs constant. Then from the CV trends, it is easier to see and interpret what is happening.

Normally, the following three design parameters need to be decided before starting the step test:

1) *Size of the step:* The amplitude must be sufficient so that excitations in the output are above the noise in the process. General guidelines in step size should be large enough to produce a 6:1 ratio of signal to noise in the key dependents.
2) *Number of steps and direction of step in the tests:* The step test should have several up steps and several down steps. Eight ups and eight down steps are normally enough to capture the dynamics.
3) *Experimental length of step:* This actually means how much time should elapse before giving another step. Different step length usually helps to identify the nonlinearity.

General guidelines for move MVs are typically 1/3, 2/3, 3/3 and 4/3 of time to steady state (TTSS), four of each.

8.2.4.2 PRBS (Pseudo Random Binary Sequence)

In the PRBS method, stepping is done automatically as per fixed plan generated by a PRBS generator. Underlying physics is based on frequency domain. Multiple MVs can be moved simultaneously and thus require less time to complete the step test. This is a potential advantage. It does not require operator intervention to carry out the step test, as moves are generated automatically. However, it is complex to interpret the results by the naked eye; PRBS software or a mathematical identifier package is required.

8.2.4.3 General Guidelines of PRBS Test

Design parameters for PRBS test includes amplitude, data collection period, switching period, and experimental length:

- *Amplitude:* Amplitude as large as can be accommodated without disturbing the process.
- *Data collection period:* Sampling period one-third to one-fifth of the fastest time constant. Set by frequency of interest.
- *Switching period:* Switching period between 1.4 and 3 times the fastest time constant.
- *Experimental length:* Experimental length between 6 and 20 times the longest settling time.

Table 8.1 Difference between Normal Step Testing and PRBS Testing

Normal Step Testing	PRBS Test
All other MVs and DVs must be held constant during test.	Operator can move any and all MVs to control the process while test is going on.
Process disturbances should be avoided and can invalidate the stepping, causing it to be repeated.	Most process disturbances do not cause any problem.
Easily interpret the step test data.	It is complex to interpret results during test.
There is more chance to disturb the process during 2 or 3 large moves.	A large number of small moves has less chance of disturbing the process.
Operator intervention is needed to give the step input during the test period.	Steps are automated. Less operator intervention is needed.
Require substantial time to complete all the steps.	Can do quicker test if all MVs can be moved at once. This is potentially the greatest advantage.

8.2.5 Difference between Normal Step Testing and PRBS Testing

Differences between normal step testing and PRBS testing are shown in Table 8.1.

8.2.6 Which One to Choose?

Nowadays, most commercial MPC packages (DMC plus, RMPCT, and SMOC etc.) have the facility to carry out PRBS testing. Underlying software has also improved over time. The software is based on making small steps on a fixed plan without operator intervention using an automatic stepping program. Two distinct advantage of PRBS testing are quicker completion of step testing and lesser intervention of operators as steps are fully automated. However, until now, most MPC vendors still preferred to carry out the normal manual step testing. This gave them more confidence, made data interpretation easier, and allowed them to take corrective actions during step testing.

General guidelines for which method to choose are as follows: If the plant type is new, and responses and process dynamics are not well known, then it is preferable to choose a normal manual step-test method. But for well-known plants, like refineries, where the process dynamics are mostly repeated and the MPC vendor has experience and confidence in expected results, and where there is a substantial benefit in time to quickly complete the test, PRBS is a better options.

8.2.7 Dos and Don'ts of Step Testing

1) Before starting step testing, ensure that basic controls are in their correct mode. In other words, mode of controllers must be in the mode they will be in after commissioning (e.g., a valve will be in manual mode if it is so as per functional design). Regulatory controller configuration must not be changed during the test.
2) Start the data collector at least 8 to 16 hours before step testing and ensure that it is working fine.
3) General rules are to make big moves with caution; the process should not go outside its allowable operating or safety limits. Move MVs typically 1/3, 2/3, 3/3, and 4/3 of time to steady state (TTSS), four of each. Some moves beyond settling time are

desirable in case the settling times were underestimated based on the pretest data. Average hold time is ½ of settling time. Long moves are valuable to identify the steady-state gain; short moves help to identify dynamics. Arrange moves into several groups throughout the plant test.

4) As explain in pretest, step size should be large enough to produce a 6:1 ratio of signal to noise in the key dependents. The dependents must "lift" out of the noise band. If this cannot be seen in the response, then no identification routine can generate good model curves.

5) Approximate time for main step test is given by
   - 12 – 16 × (number of MVs + number of DVs) × TTSS
   - Typical timing is 1 to 2 weeks for an average fractionator unit, with 4 MVs of 24 hours a day step testing.
   - It is better to calculate the approximate timing of total step testing procedure (in days or months) and communicate the plan to all concerned people to have an idea beforehand.
   - It is good practice to do more stepping than what is needed rather than less. Set expectations high for the amount of step testing.

6) Keep process around its normal range of operation during step testing.

7) Avoid CVs reaching low or high limits. Step size should be adjusted if it is so.

8) Avoid MVs reaching output limits, set point limits, etc.

9) For planned moves, only change one MV at a time. Do not control the dependent variables (CVs); allow them to drift. It will help to interpret the results later. However, sometimes you have to move more than one MV at a time to counteract something going on with the process, but try not to do this unless you must.

10) Avoid correlating moves, as moving MVs in pattern will mask the true cause/effect relationships.

11) Try to run the test in a continuous manner 24 hours/day.

12) Monitor the process carefully during step testing. Keeping an eye on data is crucial, as the data generated in step test will be used later to build the MPC data-driven model. Maintain a detailed step test log, record times of upsets and shift changes. Log abnormalities like sudden weather change, valve saturation, open bypass, instrument maintenance/malfunction, analyzer malfunction, operator changing the mode of controller by mistake, or any abnormal situations that can corrupt the data. The idea is to minimize any abnormalities or special operations during step testing. However, if any disruptions occur in spite of all precautions, then a detailed log will help to delete the outlying data during the model-building phase.

13) Working closely with operators is another crucial step. Operators are the main drivers of the step test, and their support and engagement is a prerequisite to successful step testing. Allow operators to take the initiative and lead the test. Operators should make all step changes. It is good practice to discuss the moves and their impact on CVs with the operator prior to implementation.

14) It is important to discuss the procedure with the operators so that they can understand the following:
   - It is important to move one MV at a time. Avoid correlated moves.
   - Allow CV to drift and don't try to control it. Allow sufficient time (e.g., one TTSS) as per step tests plan to hold other MVs and allow process to stabilize.

- Step size of the move should be as per step test plan. Making moves too small or making moves outside normal region of operation does not help.
- It is bad practice to stop and start a step. It is much better to go through, as then we don't have to throw away 1 TTSS data.
- Operator should have veto power to stop stepping if the operator discovers or anticipates that product can go off spec or the downstream unit may have capacity limitation.
- Operator should avoid any preventive maintenance job or any special operation like antifoam addition, heat exchanger cleaning by back flushing, or bypass opening of control valve during step testing.
- Operator should understand the implications of carrying out a good step-testing exercise.

Winning the operator confidence by properly explaining the procedure before and during step testing and engaging them actively in step testing is the winning formula of this game.

15) Try performing preliminary analysis or some model identification during step testing. This will help to identify if more or less step testing is required. If some models are found absolutely good and accurate, then the number of steps can be reduced for that particular MV-CV pair. This will give scope to increase the number of steps where bad models are identified. If a bad model is identified for particular MV-CV pair during step testing, then MPC engineers should investigate what's gone wrong. Maybe some mistakes or special operation occurred during step testing. Sometimes inclusion of some DVs into the model dramatically improves its accuracy. All the bad models should be identified quickly during step testing so that repetition of stepping can be done, if needed. Keep testing until primary models are good. Once the step test is over, then it is very difficult to convince plant management to repeat it for the sake of a bad model. It is better to identify it beforehand.

16) Note that in this preliminary analysis, a very rough assessment of model quality is made, mainly to provide guidance for testing. These models are not suitable for controller commissioning. Detailed analysis will be done after plant test.

17) Switch off the data collector at the end of step testing to avoid running out of memory in the data collector or even overwriting data at start.

8.3 Development of Step-Testing Methodology over the Years

1) The manual step-testing method introduced during 1980s has some serious drawbacks, as described below. These methods, although sufficient to get models, typically required many weeks, if not months, of bumping the process.

2) Since manual step testing requires considerable time, MPC vendors put their research efforts into developing more sophisticated automated step-testing method to reduce project time and human intervention.

3) Development of new step-testing methods started in the early 1990s and is still evolving. New research in this area has enabled operators to reduce MPC project time threefold over the last three decades.

4) Currently, intelligent stepping software is available to commercial MPC vendors (Honeywell, Aspentech etc.) that incorporates decades of experience. The software automatically performs many of the tasks previously done manually, such as stepping the process, collecting data, identifying models, and validating models. The software even allows conducting these tasks while an MPC application is controlling the process. This enables step testing with less impact on the process operation and stepping while the process is being controlled. This modern stepping software is opening the door to implementing MPC on processes previously considered inapplicable to MPC. It also affords itself to easing the burden of maintaining existing MPC applications.

5) Step-testing methods have evolved over 40 years. A brief history and milestones achieved are given here (John 2014):

- Manual open-loop step-testing methods were used from 1970 to 1990. It is still practiced in industry in large numbers and is considered a proven methodology.
- In 1990, open-loop sequential automated step-testing methods were first deployed.
- 2000 marked the development of open-loop simultaneous automated step-testing methods.
- In 2005, open-loop automated step testing was deployed in conjunction with automated model identification methods.
- Also in 2005, closed-loop automated step testing in conjunction with an online MPC and automated model identification methods were deployed to maintain existing MPC applications.
- 2010 saw enhanced closed-loop automated step testing in conjunction with an online MPC and automated model identification methods.

The main goal has always been to acquire quality models as quickly and with as little disruption to the process as possible. Progress over the past decades has been significant, with higher-quality models acquired amid less disruption to the process and in shorter time frames.

Reference

John A. Escarcega, (June 2014), White paper on "An Evolution of Step Testing and its Impact on Model Predictive Control Project Times," Honeywell International Inc., Houston.

9

Model Building and System Identification

9.1   Introduction to Model Building

At the completion of plant step test, the plant data are available. The next step is to build an accurate model (MV-CV and DV-CV relationship), which will be used for dynamic prediction of the system. Before a MPC can be properly configured, a model of the process must be identified. Control engineers need to be concerned with both the steady-state performance and the dynamic response of the process. It is the dynamic response of a process that has a fundamental impact on the design of the control functions.

Dynamic response implies a change of a variable with respect to time. The variable could be temperature, pressure, composition, level, etc. The change could be caused by a controlled disturbance, such as a change in feed rate, or an uncontrolled disturbance, such as change in feed quality. Dynamic models are mathematical representations of the transient relationships between the process output variables and the process input variables. Although process models can sometimes be derived from historical plant data, they are typically obtained from plant step tests.

Here are some commonly used concepts in model identification:

- *What do you mean by a model?* In the context of MPC, it is an estimation of the true behavior of a dynamical system (chemical or physical process). Based on our process knowledge and observations of the system, model means a mathematical expression that can predict dependent variables (CVs) in terms of independent variables (MV or DV). It is extremely important to discover the cause-and-effect structure amongst inputs and outputs as well as the individual model parameters.
- *What is a dynamical system?* It is a system in which variables of different kinds interact and produce observable signals that evolve in time. In the context of MPC, a dynamic system is any process unit in the plant.
- *How to build a model of the process?* The first principal base model is this: Models can be built with fundamental mass, energy balances, and generating partial differential equation governing them. However, modern chemical processes in commercial plants are too complex and time consuming to be modeled by first principals. In those cases, the system identification technique is commonly used.
- *What is system identification?* System identification is a technique that allows users to build mathematical models of a dynamic system based on measured data.

Multivariable Predictive Control: Applications in Industry, First Edition. Sandip Kumar Lahiri.
© 2017 John Wiley & Sons Ltd. Published 2017 by John Wiley & Sons Ltd.

- *How is that done?* Essentially, by adjusting parameters within a given model until its output coincides with the measured output.
- *What is a good model?* In the context of MPC, a good model is one that ensures good controller performance. Recipe for good models are listed below:
  - Process knowledge—good models captures the true cause-and-effect structure between MV and CV.
  - Informative step test data—good models uses good informative data, which means data contains good signal/noise ratio; uncorrelated input data; and data pertaining to the operating regime of the controller.
  - Step test done at stable process—step test was done as a stable process without unwanted disturbances and with good tuning of base-level regulatory control.
  - Understanding the controller objectives: Since the litmus test of a good model is to ensure good controller performance, it is important to know both operational and economical controller objectives.
- *How do you know if the model is any good?* A good test is to take a close look at the model's output compared to the measured one on a data set that wasn't used for the fit (validation data). It is also valuable to look at what the model couldn't reproduce in the data (the residuals). This should not be correlated with other available information, such as the system's input.

Model sets or *model structures* are families of models with adjustable parameters. *Parameter estimation* amounts to finding the "best" values of these parameters. The *system identification* problem amounts to finding both a good model structure and good numerical values of its parameters.

Parametric identification methods are techniques to estimate parameters in given model structures. Basically, it is a matter of finding (by numerical search) those numerical values of the parameters that give the best agreement between the model's (simulated or predicted) output and the measured one.

Nonparametric identification methods are techniques to estimate model behavior without necessarily using a given parametrized model set.

Typical nonparametric methods include *correlation analysis*, which estimates a system's impulse response, and *spectral analysis*, which estimates a system's frequency response.

Research provides an excellent technology overview (DMC Corp. 1994; Honeywell Inc. 1995).

9.2 Key Issues in Model Identifications

Chemical process industry (CPI) processes can be characterized as (1) large-scale and complex, (2) dominant slow dynamics, and (3) high-level disturbances. It requires special attention in CPI for model identification. This section will describe the four problems of identification: test design, parameter estimation, model structure and order selection, and model validation.

9.2.1 Identification Test

In a traditional identification test, each MV is stepped manually and each MV is tested separately after each other. All the CVs are in open-loop operation. The average step length is related to the estimated settling time of the process. The test is carried out around the clock, and it will cost 15 to 20 days to test a large unit such as a crude unit and a FCCU. This approach has been very successful in the last 20 years. The advantage of this test method is that control engineers can watch many step responses during the tests and can learn about the process behavior in an intuitive manner. The problems with single variable step tests are:

- High cost in time and in manpower.
- The data from a single variable test may not contain good information about the multivariable character of the process (ratios between different models) and step signals do not provide sufficient excitement of the dynamic character of the process.
- An open-loop test may disturb unit operation.

Using automatic multivariable closed-loop testing can solve these problems. There are many advantages of a multivariable closed-loop test:

- *Less disturbance to unit operation.* In a closed-loop test, the controller will help to keep the CVs within their operational limits.
- *Easier to carry out.* In an automatic multivariable closed-loop test, much less engineer or operator intervention is needed. Night shifts may be avoided.
- *Better model for control.* This can be explained in several ways. Under the same CV variance constraints, the model from a closed-loop test data will have higher control performance than the model from an open loop test.

Advantages of identification techniques in a closed loop are explained in detail (Gustavsson 1977; Hjalmarsson 1996; Snow 2001).

9.2.2 Model Structure and Parameter Estimation

In traditional MPC identification, first an MIMO FIR model is used to estimate, using least-squares method. This often results in a model with nonsmooth step responses. Model reduction or smoothing techniques are used to obtain smooth model responses. Optionally, SISO parametric models are estimated using data slices that only involve the movements of one MV-CV pair. This is not always feasible due to high-level disturbances and multiple movements of the MVs.

The following models/methods are common in identification literature:

- FIR (finite impulse response) model
- ARX (AutoRgressive with eXternal input) model, or, least-squares model
- Output error (OE) model
- ARMAX (AutoRegressive Moving Average with eXternal input) model
- Box-Jenkins model

These are special cases of the more general prediction error model family (Ljung, 1985; 1999). The model parameters are determined by minimizing the sum of squares of the prediction error. In literature, ARX, OE, ARMAX and Box-Jenkins models are called parametric models and the FIR model is called a nonparametric model.

Subspace method of parameter estimation has been proposed and studied in the literature; see van Overschee and de Moor (1994), Verhaegen (1994), and Larimore (1990). Subspace methods estimate a state space model of a multivariable process directly from input/output data.

For closed-loop identification, the choice of model structure (or estimation method) depends on three often-conflicting issues:

1) The compactness of the model
2) The numerical complexity in parameter estimation
3) The consistency of the model in closed-loop identification

When noisy data are used in identification, a more compact model will be more accurate, provided that the parameter estimation algorithm converges to a global minimum and the model order is selected properly. In general, a model structure or an estimation method that includes a disturbance model will be more accurate than a method without the disturbance model. Moreover, a model with a disturbance model will give a consistent estimate for closed-loop data, meaning that the effect of the disturbance will decrease when test time increases; whereas a model without a disturbance model will deliver a biased estimate when using closed-loop data.

However, a more compact model needs more complex parameter estimation algorithms. To estimate OE models, Box-Jenkins models, and ARMAX models, nonlinear optimization routines are needed that often suffer from local minima and convergence problems. In FIR and ARX models, the error term is linear in the parameters. Due to this property, a linear least-squares method can be used in parameter estimation that is numerically simple and reliable. This partly explains why the FIR model is often used in industrial identification. The subspace methods are exceptional: They estimate a parametric model and they are numerically efficient. The main part of a subspace method consists of matrix singular value decomposition (SVD) and linear least-squares estimation, which are numerically simple and reliable. To summarize the discussion, the Figure 9.1 compares the advantages and disadvantages of various model structures or parameter estimation.

For multivariable processes, model parametrization (SISO, MISO or MIMO) will also play a role in determining model accuracy for control; see Zhu and Butoyi (2002).

9.2.3 Order Selection

In traditional MPC identification, for an FIR model, the estimated settling time is used as the model length or "order" of the model. This is theoretically simple, but it is not easy to use in practice. A phenomenon that often occurs is that, when the disturbance

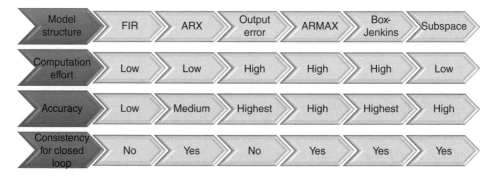

Figure 9.1 Advantages and disadvantages of various model structures

level is high, the model gains will change when the model length changes. For SISO model estimation, usually first order or second order plus delay models are tried and the choice is made based on simulation and/or process knowledge.

For the purpose of control, it is most important to select the model order so that the process model is most accurate. In the time domain, this requires that the simulation error, or, output error of the model be minimal; in the frequency domain, this requires that the total error is minimal. See Zhu (2001) for more details.

9.2.4 Model Validation

In traditional MPC identification, the model validation and selection are done based on process knowledge on the gains and on the fits between simulated CVs and their measurements.

The goal of model validation is to test whether the model is good enough for its purpose and to provide advice for possible reidentification if the identified model is not valid for its intended use. Simulation approach is very questionable for multivariable closed-loop test data because a good fit of a CV cannot guarantee that models from all MVs are equally good and a poor fit of a CV does not imply that all the models for that CV are bad. A more useful model validation method should provide information of model accuracies of each transfer functions and relationship between model accuracies and test variables such as signal amplitudes and test time.

System identification remains both an art and a science. The *science* is concerned with parameter estimation; the *art* is usually concerned with determining structure/order, the excitation requirements, and accuracy. System identification involves two steps:

1) A sequence for exciting the process to be modeled is specified. A family of candidate models is then proposed. After this, a representative member of this family is selected. This is the *art* and is often problem specific.
2) The second step is the *science*. This step is a parameter-estimation problem. Parameter estimation is basically the determination of the *best* set of candidate model coefficients such that the model represents the causal input/output relationships.

9.3 The Basic Steps of System Identification

The system identification problem is to estimate a model of a system based on observed input-output data. Several ways to describe a system and to estimate such descriptions exist. This section gives a brief account of the most important approaches.

The procedure to determine a model of a dynamical system from observed input-output data involves three basic ingredients:

1) The input-output data
2) A set of candidate models (the model structure)
3) A criterion to select a particular model in the set, based on the information in the data (the identification method)

The identification process amounts to repeatedly selecting a model structure, computing the best model in the structure, and evaluating this model's properties to see if they are satisfactory.

The cycle can be itemized as follows:

1) Design an experiment and collect input-output data from the process to be identified.
2) Examine the data. Polish it so as to remove trends and outliers, and select useful portions of the original data. Possibly apply filtering to enhance important frequency ranges.
3) Select and define a model structure (a set of candidate system descriptions) within which a model is to be found.
4) Compute the best model in the model structure according to the input-output data and a given criterion of fit.
5) Examine the obtained model's properties.
6) If the model is good enough, then stop; otherwise, go back to Step 3 to try another model set. Possibly also try other estimation methods (Step 4) or work further on the input-output data (Steps 1 and 2).

The system identification or model building part of MPC offline software offers several functions for each of these steps. For Step 2, there are routines to plot data, filter data, and remove trends in data, as well as to resample and reconstruct missing data.

For Step 3, the MPC software offers a variety of nonparametric models, as well as all the most common black-box input-output and state-space structures, and also general tailor-made linear state-space models in discrete and continuous time.

For Step 4, general prediction error (maximum likelihood) methods, as well as instrumental variable methods and sub-space methods are offered for parametric models, while basic correlation and spectral analysis methods are used for nonparametric model structures.

To examine models in Step 5, many functions allow the computation and presentation of frequency functions and poles and zeros, as well as simulation and prediction using the model.

Figure 9.2 shows a simple flowchart of identification process.

9.3.1 Step 0: Experimental Design and Execution

System identification is about building mathematical model of dynamic systems based on measured data. This process includes the following steps, which have already been discussed in previous chapters:

1) Understand the key dependent and independent variables in the process. Identify CV-MV-DV list of the process.
2) Preliminary process test (pre-stepping) to identify and remove all the potential barriers for a good step test.
3) Properly plan and execute step test and/or PRBS test under supervision of MPC engineer.
4) Generate good-quality step test data (high signal to noise ratio) with proper log book of all the activities.
5) Be aware of the key disturbances and investigate potential means to eliminate them or their effect (i.e., at source, or as a part of current strategy)
6) Make sure you are clear on the economic drivers of the process in terms of feedstocks and products.

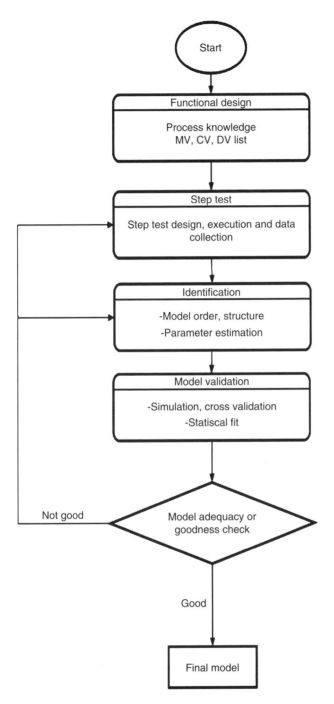

Figure 9.2 Flowchart of identification process

9.3.2 Step 1: Plan the Case that Needs to Be Modeled

9.3.2.1 Action 1

At this stage, a preliminary controller design case (i.e., MV-CV, DV-CV list) has already been identified in the step test procedure. In other words, the MV, CV and DV that need to be included in the controller design have already been determined. The purpose of this step is to reconfirm that list or modify that list, as deemed fit by analyzing the step test data. Often during model identification step, this preliminary design is modified. As step test data contain information and knowledge of the process, it may possible to discover new information and insights about the process interaction at this stage, which may have been earlier overlooked or not explored.

A thorough inspection of step test data is required in this step. From the analysis of data, there are several reasons that the design may be modified at this point. Some possible reasons (but not limited to) are given:

- A MV in the controller design is found to have a relatively minor effect on the system. It is either changed to a disturbance variable or dropped from the controller entirely.
- During step test, a CV not included in the controller design is found to be a constraint on the system. That is, a valve may open more that 95 percent during the step test. This variable (say valve opening) is included in the controller as a CV.
- A DV already in the controller design is found to have only a minor effect on the system. In this case, it may be dropped from the controller.
- A potential disturbance that was NOT included in the controller design is found to have a significant effect on the system, and is thereby included in the controller.
- During the model identification phase, a new way to calculate an inferred property calculation is discovered. This new inferred property is added to the controller design as a CV.

This step is ongoing, and controller design can be modified at any stage of model identification step.

9.3.2.2 Action 2

Using the preliminary controller design so far, an expectation matrix must be prepared now, as shown in Figure 8.1 in chapter 8. This constitutes the plan for the analysis cases. The matrix is made up of all independent and dependent variable that are to be included in the controller. Once this matrix is filled in, it becomes apparent that it can be divided into multiple groups for analysis.

9.3.3 Step 2: Identify Good Slices of Data

As the MPC models are built from the data, models are as good or as bad as the quality of data. It is an extremely important part of the model identification process to identify the valid, good, time slices of data that should be used for model building. Bad data should be excluded from the analysis as it will lead to incorrect model.

The logbook kept during the plant test is an important source of information about which data must be excluded from the analysis. Pump failure, antifoam addition, and exchangers being taken out of service are events that might invalidate data.

In addition to the logbook, a thorough analysis of the data should be made by plotting them in one window using plot facility of the MPC offline software.

9.3.3.1 Looking at the Data

Plot the data. Look at them carefully. Try to see the dynamics with your own eyes. Can you see the effects in the outputs of the changes in the input? Can you see nonlinear effects, like different responses at different levels, or different responses to a step up and a step down? Are there portions of the data that appear to be "messy" or carry no information? Use this insight to select portions of the data for estimation and validation purposes.

Do physical levels play a role in your model? If not, detrend the data by removing their mean values. The models will then describe how changes in the input give changes in output, but not explain the actual levels of the signals. This is the normal situation. The default situation, with good data, is that you detrend by removing means, and then select the first half or so of the data record for estimation purposes, and use the remaining data for validation.

Here are practical guidelines regarding how to inspect and investigate data:

- The first points to inspect are those data associated with regulatory controller encompassed by MPC. They may be MV or DV. If any of these controllers goes off control during the test, whether by the operator changing the mode or the valve becoming saturated, the data recorded at that time are invalid.
- If an upstream level control valve becomes saturated (or operator put it in manual mode), the changes in the upstream section are no longer passed to the downstream towers. The system behaves quite differently. These data must not be used.
- Other events that may invalidate data are major upsets during a step test, controllers not at their desired mode of operation, process value freezed, or data collection systems not working for some time. Some special operations during step testing include pump changeover, and heat exchanger cleaning, compressor surging.
- Note that bad data in MV or DV have broader ramifications than bad data in CV.

9.3.4 Step 3: Pre-Processing of Data

Data gathered from process industries normally involves the following critical characteristics:

- Data contain spike or very high value.
- Data contain noise due to process itself or from measuring transmitters.
- Data may contain missing value or value may be freezed.

The following tasks are performed in this step;

- Data cleaning
- Outlier detection and replacement
- Handling of missing data
- Selection of relevant variables (i.e., feature selection)

Normally, all the MPC offline software has inbuilt capability to do all the data preprocessing. A detailed procedure regarding data cleaning and outlier detection is given in Chapter 10.

9.3.5 Step 4: Identification of Model Curve

A fundamental problem in system identification is the choice of the nature of the model, which should be used for the system. In that respect, system identification still remains both an art and a science. The science is concerned with parameter estimation, while the art is concerned with overcoming the following problems:

- Determining structure and/or order
- Excitation requirements
- Accuracy

System identification requires two steps. First, a family of candidate models is decided on (structure selection). Then the particular member of this family that satisfactorily describes the observed data is selected (parametric estimation).

Different MPC vendors are handling this model-building procedure in different styles and techniques. MPC technology provided by different vendors has different identification techniques. Choices of model available in their software are also different. For example, DMC plus has two type of modeling choices, such as FIR and subspace identification. RMPCT has FIR and prediction error method (PEM) –based modeling approach. Based on the model available in their software, different vendors follow different identification techniques.

There are two major approaches for identification:

1) Hybrid approach (mainly followed by RMPCT and SMOC with some variations)
2) Direct modeling approach (followed by DMC plus with some variations)

Based on industrial experience and model complexity, a hybrid approach is recommended.

9.3.5.1 Hybrid Approach to System Identification

In this type of system identification approach, the system is identified in three phases. The generated model passes through three phases to arrive at a final model.

First Phase: Fitting FIR or PEM models

In first phase, a Finite Impulse Response FIR model is identified and fitted with step test data. Additionally, sometimes if FIR model found unsatisfactory, then a black-box models based on PEM is fitted with step test data.

Fitting FIR Models

FIR models are based on raw plant data and have these characteristics:

- Solutions result in an unbiased estimate (when plant tests are conducted in the open loop with a properly designed signal).
- Models are structure free.
- Solutions are extremely fast and exceedingly robust.
- Solutions result in potentially high-order, high-variance estimates (damping these estimates with weights or smoothing will result in biased estimates, and in addition will cause problems in the calculation of confidence limits).

Fitting PEM Models

PEM models are based on raw plant data and have these characteristics:

- Solutions result in a consistent estimate (when structure is compatible with the process and when the procedure converges to the global minimum). This is true even under closed-loop operation.
- For Gaussian disturbances, the procedure results in a minimum variance estimate.
- Goal is a one-step (load and go) operation.
- Relatively low-order models imply minimal information loss under segmented or contaminated data conditions.
- Solutions are obtained numerically.
- Nonlinear search is required in which convergence *cannot* be guaranteed.

Usually, the PEM structure supports the following standard forms: FIR, ARX, ARMA, ARMAX, ARIMA(X), ARARMAX, BJ, and OE.

Second phase: Fitting Parametric Models
In second phase, a low-order parametric model is developed by taking either FIR or PEM results as input. The purpose of the parametric model is to take the FIR or PEM model, and fit it with a parametric model that reduces or eliminates the variance. In addition, the parametric models provide an extremely effective mechanism for model order reduction that is easily configured by the user.

At this step:

- Low-order parametric models are developed.
- FIR or PEM models are used to get excellent initial guesses for the parametric models. This includes an initial guess for the dead time.
- Normally, in the identification software, choices of several parametric models are available. Commonly available parametric models are given below:
  - Laplace domain (requires nonlinear search)
  - Discrete domain
  - ARX with prefiltering
  - Output error (requires nonlinear search)
- Model determination is highly robust and rapidly convergent since structure here is not a factor and the estimation is done on the FIR/PEM results not on the raw data.
- All parametric models are ultimately converted to Laplace domain form irrespective of the original form.

Final Phase: Fitting Final System Models
In the third and final phase, the best parametric models are selected based on their statistical fitting performance on validation test set data. Parametric models are automatically selected for the final system model. Based on the raw data (cross validation if selected), the parametric models with the best long-term open-loop prediction performance are automatically selected for use in the final system model. Models can also be automatically eliminated based on FIR statistics. Reduced order model verification and selection are done based on process knowledge on the gains and on the fits between simulated CVs and their measurements.

9.3.5.2 Direct Modeling Approach of System Identification
Two modeling techniques are supported in this type of approach—namely FIR and subspace identification.

FIR Identification

- During a step test period, collect time-stamped process data from the system for all independent and dependent variables.
- From the collected data, calculate δCV and ΔMV.
- Solve the control equation, δCV = A * ΔMV, for the unit step response coefficients, the A matrix.
- To make the modeling technique more robust, we need a solution method that is tolerant of multiple moves within a single TSS (i.e., simultaneous independent variable moves). Benefit of simultaneous solution allows for changes in more than one independent variable at a time during the test. This means shorter step test duration.
- For simultaneous identification, convert the control equation to a residual form δCV − A * ΔMV = R
- Now, we can use a *least squares* regression method to solve for the response coefficients while minimizing R^2.
- Calculate the sum of the squared residual terms:

$$r^T r = r_1^2 + r_2^2 + r_3^2 + r_4^2 + r_5^2 + r_6^2 + r_7^2 + r_8^2 + r_9^2$$
$$= \sum r_i^2$$
$$= \text{Sum of Squared Residuals}$$

- The advantage of this type of FIR modeling is that model order/form is not a constraint on the identification problem using FIR models. All dynamic information in the model is based on actual process information. Any type of response can be modeled explicitly.
- One of the drawbacks is that process information must provide not only dynamic information, but also form. Additional process data are required to support this nonordered model.
- For an FIR identification run, the following parameters are required:
  1) Time to steady state (or settling time)—This is the expected maximum length of the steady state model in minutes.
  2) Number of model coefficients—The number of points used to draw the model curve. More coefficients are required to model faster responses.
  3) Smoothing factor—The parameter used to smooth the data and apply the penalty for change between successive FIR model coefficients.

9.3.5.3 Subspace Identification

Subspace identification is a new (compared to traditional transfer function) parametric identification method that uses a state-space model to represent process internal dynamic relationships. The identification algorithm is able to capture both low-frequency and high-frequency (i.e., both low-order and higher-order) dynamics. Its superiority is claimed by the fact that it can model many difficult dynamic processes, such as inverse response (nonminimum phase behavior), stiff dynamics, nonstationary noise, unknown dead times, and integrating and unstable processes.

The subspace identification technology is particularly suitable to multiple-input, multiple-output (MIMO) identification, which meets the needs of MPC controllers very well. Applying MIMO identification can achieve minimal parameterization and improve identification efficiency.

- From its computational mechanism, it is known that the subspace identification algorithm is noniterative, and it uses numerically stable linear algebra.
- As a result, no convergence problems arise, and no parametric nonlinear programming is required. The algorithm is expected to offer robust identification and more accurate models than other methods.

Subspace Identification Parameters
- Time to steady state: For subspace identification, the time to steady state is the expected settling time of the process.
- Maximum order: In subspace identification, an order is a dynamic state used to capture the dynamics of a process. The maximum order specifies the maximum of these dynamic states that may be considered by the subspace identification algorithm in its search for an optimal model order.
- Subspace identification takes more computation time than the FIR identification, in particular for models with a large number of independent and dependent variables. This occurs because the subspace identification does a true multi-input, multi-output (MIMO) model identification that needs more intensive computation than the multi-input, single-output (MISO) FIR identification. The benefit is that a true MIMO model has less uncertainty, and is statistically more accurate.

The subspace identification technology has many attractive features and advantages over other identification methods, but it is comparatively new, especially for industrial practice (say, last 5 to 10 years). On the other hand, the nonparametric FIR identification algorithm used with MPC model is a very well developed, field-proven modeling tool used for over 35 years in the industry. It is advisable to run both FIR and subspace trials in the same case, and compare the results and then select the final model.

9.3.5.4 Detailed Steps of Implementations
Both of the previous approaches of system identification have some strength and weakness, and their effectiveness varies case to case.

The following section summarizes how to implement the hybrid approach in actual practice. The implementation steps of the hybrid approach are summarized below:

1) *Set up overall model for FIR by specifying maximum settling time and number of co-efficient.* In the first step of system identification it is always recommended to use FIR curve as it is robust, easy to interpret, generally use with final step tests and only settling time need to specify as priori. The only true independent variable is the settling time. While precise values are not important, a range from low (50 percent smaller than the smallest) to large (50 percent larger than the largest) will give good results.

2) *Generate FIR models with step test data and generate step responses curve.* All step responses (one for each settling time) for each sub-model are plotted on the same plot. Similar curves for two or more settling times indicate reasonable models. Separation of the curves at longer settling times indicate that too long a settling time was used.

- Step responses based on FIR coefficients can have a large variance. The variance is reduced in the parametric sweep. No special processing to smooth the FIR models is performed because useful information is available in the raw models. In most cases, filtering is not recommended.
- Random looking responses or responses that change sign in the mid to low frequency range indicate that there is no model.

3) *Judge the quality of model by CV to MV/DV correlation and other statistical method.* CV to MV/DV cross correlation useful for detecting closed loop step testing, Strong versus weak correlation. Use different statistical analysis (e.g., Non-null hypothesis test) for early assessment of model quality (very useful).

4) *Fit parametric model to the FIR curves.* Use Laplace transform or ARX method to fit. Use Laplace transform or ARX method to fit.

5) *Compare actual vs. model predicted curve.* See statistical fitting parameters.

Parametric Model Fit Options

Parameter estimation is basically the determination of the 'best' set of candidate model coefficients such that the model represents the observed data in a desirable fashion.

Primary reason for the parametric model fit is to reduce model variance. Additional benefits, models have the minimum number of model parameters to represent the dynamic behavior. For example, a process can represent a first-order dead time and lag response with a 60 FIR parameters or with three parametric parameters. If it is desirable or necessary to edit the identified models, this is much easier with parametric models. Anyone can modify the model very easily on the fly.

Model Selection

Individual parametric submodels are selected on the basis of minimum prediction error with respect to the FIR step response models not the raw data. Choices for Selection is between Laplace, ARX, and/or Output Error.

One continuous time model will exist for each submodel for each separate settling time.

Evaluate Fitting Statistics

There are many statistical methods available to see the goodness of fitted parametric model. These statistical indicator shows whether the estimated regression model is statistically significant.

Model validation is performed on full order models in terms of; confidence limits, noise bounds, null hypothesis tests and step response sensitivities. Model rankings are automatically determined based on the above-mentioned statistics.

9.3.6 Step 5: Select Final Model

Final stage in the identification process is the selection of the best models. As discussed earlier, it is necessary to have a reasonable estimate of the process settling time before an FIR model can be fitted to the data. To overcome uncertainties of determining the optimal settling time, typical to run a number of alternative cases of the FIR identification, each with its own settling time (e.g., three cases 60-, 90-, and 120-minute settling time). This also accommodates the fact that different CVs have different selling times from each other. For each FIR identification case, a parametric model is fitted to each

CV, MV (or DV) pair. Finally, reduced-order model verification and selection are done based on process knowledge on the gains and on the fits between simulated CVs and their measurements.

It is extremely important to judge whether a developed model is good enough to deploy in an online control system. How can you know the developed model is good enough?

1) Check that step responses/gains make sense process-wise.
2) Visually inspect predicted vs. actual process output.
3) Check F-statistics (not always a good indication).
4) Check confidence intervals.
5) Conduct correlation analysis.
6) Check the performance on other validation data set.

9.4 Model Structures

System identification is not necessarily about fitting data, but rather, about finding the causal relationships between the inputs u and the outputs y shown in the Figure 9.3. This is to be accomplished in spite of the unmeasured disturbances v (refer Figure 9.3). While in some instances it may be desirable to obtain a disturbance model H (z), the ultimate objective is to obtain the rational transfer function matrix G (z) whether H (z) is determined or not. In some cases, a "good" model may yield a poor fit of the data while in others a "poor" model may yield a good fit of the data. It is the objective of the identifier to not only extract as much useful information out of the data as possible, but to also indicate whether in fact the models obtained are useful for the purposes of process control.

Modeling is accomplished in two distinct phases. In the first phase models are used to regress the data. At this level, two different model types are supported, already

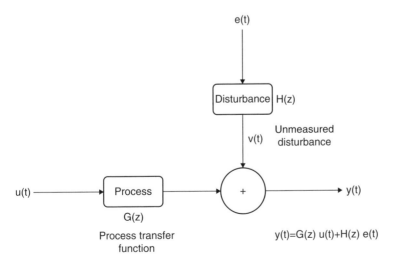

Figure 9.3 System identification structure

introduced: the finite impulse response (FIR) model and the generalized black-box model based on the prediction error method (PEM). Rather than a single model, PEM represents virtually all of the standard discrete-time polynomial models.

In the second phase, models are used for order reduction and to reduce or eliminate variance. Inputs here are the models obtained in the first phase based on regressed data. At this level, three distinct types of models are supported. Two are discrete or z-domain models and one is a continuous or s-domain model. The discrete models are ARX and output error (OE). The ARX model is actually a prefiltered ARX, where the prefilter is used to weight the low frequency fit. The order of the discrete time model can be defined by the user and is unrestricted. The s-domain model has a fixed structure form in which the order (up to three) is automatically determined. Discrete models in this phase are ultimately converted to the s-domain before they are saved.

9.4.1 FIR Models

With the proper formulation, the FIR approach can be an exceptionally effective estimator. APC identifier uses the FIR model as its base form.

9.4.1.1 FIR Structures

The general FIR structure used is given by:

$$y_t^i = \left(p_0^{i,1} u_t^1 + p_1^{i,1} u_{t-1}^1 + p_2^{i,1} u_{t-2}^1 + \ldots + p_{n1}^{i,1} u_{t-n1}^1 \right) + \left(p_0^{i,2} u_t^2 + p_1^{i,2} u_{t-1}^2 + p_2^{i,2} u_{t-2}^2 + \ldots + p_{n2}^{i,2} u_{t-n2}^2 \right)$$
$$+ \ldots + \left(p_0^{i,m} u_t^m + p_1^{i,m} u_{t-1}^m + p_2^{i,m} u_{t-2}^m + \ldots + p_{nm}^{i,m} u_{t-nm}^m \right)$$

This expression corresponds to the positional form of the FIR model. There are no inherent limitations imposed on the structure of this model, and as such, each sub model (i, j element) is free to have as many coefficients as necessary to adequately capture the observed response.

To obtain the velocity form of the model, all dependent and independent variables in the above expression are differenced in time. Differencing is invoked at the discrete sample rate of the model, which is in general different from the sampling time of the data.

In terms of the general ID structure presented previously, the FIR model can be written conveniently as

$$y(t) = \sum_i A_i\left(z^{-1}\right) u_i(t) + v(t) + \alpha$$
$$G_i\left(z^{-1}\right) = A_i\left(z^{-1}\right)$$

while the predictor in positional and velocity form respectively is given by:

$$\hat{y}(t) = \sum_i A_i\left(z^{-1}\right) u_i(t) + \alpha = \sum_i \left(a_0 u(t) + a_1 u(t-1) + \ldots + a_{nb} u(t-nb) \right)_i + \alpha$$
$$\Delta \hat{y}(t) = \sum_i A_i\left(z^{-1}\right) \Delta u_i(t) + \alpha$$
$$\Delta y(t) = y(t) - y(t-1)$$
$$\Delta u_i(t) = u_i(t) - u_i(t-1)$$

9.4.2 Prediction Error Models (PEM Models)

This model form encompasses virtually all of the polynomial black-box models. In its full form, this model can be used in both the open and closed loop and under reasonable conditions is a consistent estimator.

While PEM models can be used in a general setting, due to computation, structure, and convergence limitations, they can be less effective from a practical perspective than the standard FIR approach.

9.4.2.1 PEM Structures

The general PEM structure used is given by:

$$A(z^{-1})y(t) = \sum \left[\frac{P(z^{-1})}{Q(z^{-1})} u(t-d) \right] + \frac{R(z^{-1})}{S(z^{-1})} e(t) + \alpha$$

$$G(z^{-1}) = \frac{P(z^{-1})}{Q(z^{-1})A(z^{-1})}$$

$$H(z^{-1}) = \frac{R(z^{-1})}{S(z^{-1})A(z^{-1})}$$

Based on this model, the corresponding predictor is:

$$\hat{y}(t) = \frac{S(z^{-1})}{R(z^{-1})} \sum \left[\frac{P(z^{-1})}{Q(z^{-1})} u(t-d) \right] + \left[1 - \frac{D(z^{-1})A(z^{-1})}{R(z^{-1})} \right] y + \frac{S(z^{-1})}{R(z^{-1})} \alpha$$

Based on the above expression, the prediction error, defined as $\varepsilon(t) = y(t) - \hat{y}(t)$, can be written as:

$$\varepsilon(t) = \frac{S(z^{-1})}{R(z^{-1})} \left[A(z^{-1})y(t) - \sum \left[\frac{P(z^{-1})}{Q(z^{-1})} u(t-d) \right] - \alpha \right]$$

The polynomials in the above expressions have the following form:

$$A(z^{-1}) = 1 + a_1 z^{-1} + a_2 z^{-2} + \ldots + a_{na} z^{-na}$$
$$P_i(z^{-1}) = \left(p_0 + p_1 z^{-1} + p_2 z^{-2} + \ldots + b_{nb} z^{-nb} \right)_i$$
$$Q_i(z^{-1}) = \left(q_0 + q_1 z^{-1} + q_2 z^{-2} + \ldots + q_{nq} z^{-nq} \right)_i$$
$$R(z^{-1}) = 1 + r_1 z^{-1} + r_2 z^{-2} + \ldots + r_{nr} z^{-nr}$$
$$S(z^{-1}) = 1 + s_1 z^{-1} + s_2 z^{-2} + \ldots + s_{ns} z^{-ns}$$

In commercial MPC software usually PEM structure supports the following standard forms:

FIR, ARX, ARMA, ARMAX, ARIMA(X), ARARMAX, BJ, OE

9.4.3 Model for Order and Variance Reduction

The following paragraphs illustrate those models that are used for order reduction and variance reduction/elimination. The discrete time models presented below correspond to structures contained in the PEM model. They should, however, not be confused with the PEM approach. The models presented here are much simpler by design and, as such, have their own solution methodology.

9.4.3.1 ARX Parametric Models (Discrete Time)

Parametric models are used for model order reduction and to remove the variance found in the models regressed from raw data. While standard low-order ARX models are typically inadequate due to biased estimates, the prefiltered form used in the APC identifier automatically weights the low frequency fit and hence, results in high-quality models.

The general form of this model is given by:

$$P(z)y'(t) = B(z)u'(t-d) + e(t)$$

The resulting transfer function takes the form:

$$T(z) = \frac{\left(b_1 z^{-1} + b_2 z^{-2} + \ldots + b_n z^{-n}\right) z^{-d}}{\left(1 + p_1 z^{-1} + \ldots + p_n z^{-n}\right)}$$

In the above expressions the prime denotes a prefiltered value while n and d correspond to the order and delay of the subprocess, respectively.

9.4.3.2 Output Error Models (Discrete Time)

In addition to the ARX form, the identifier also generates discrete time models with an output error structure. The general form of these models is given by:

$$w_t + f_1 w_{t-1} + f_2 w_{t-2} + \ldots + f_n w_{t-n}$$
$$= b_1 u_{t-1-d} + b_2 u_{t-2-d} + \ldots + b_n u_{t-n-d}$$
$$y_t = w_t + e_t$$

Close inspection of these expressions shows that the output y does not appear in the regression matrix for this model. Consequently, this structure results in an unbiased estimate if the input u is persistently exciting. The above expressions can be conveniently written as:

$$y(t) = \frac{B(z)}{F(z)} u(t-d) + e(t)$$

The resulting transfer function takes this form:

$$T(z) = \frac{\left(b_1 z^{-1} + b_2 z^{-2} + \ldots + b_n z^{-n}\right) z^{-d}}{\left(1 + f_1 z^{-1} + \ldots + f_n z^{-n}\right)}$$

While the output error model has the desirable feature of being unbiased even without prefiltering, this structure requires that the estimation parameters appear in the regression matrix. Consequently, the estimation problem becomes nonlinear. This implies that more computational effort is required for the output error solution than is required for the ARX solution.

9.4.3.3 Laplace Domain Parametric Models

It is also possible to generate parametric models directly in the Laplace domain. The general Laplace domain form is given by:

$$T(s) = \frac{k(\tau s + 1)e^{-ds}}{s(\tau_1 s + 1)(\tau_2 s + 1)}$$

This model is guaranteed to be overdamped and open-loop stable.

9.4.3.4 Final Model Form

All models are ultimately saved in Laplace form. Discrete models are automatically converted from the z to s domain. The form of the saved models is:

$$T(s) = \frac{k(b_{n-1}s^{n-1} + \ldots + b_1 s + 1)e^{-ds}}{s(a_n s^n + a_{n-1}s^{n-1} + \ldots + a_1 s + 1)}$$

The lead s in the denominator is present only if the subprocess contains an integrator. In this case, the k refers to the integration rate.

9.4.4 State-Space Models

State-space models are common representations of dynamical models. They describe the same type of linear difference relationship between the inputs and the outputs as in the ARX model, but they are rearranged so that only one delay is used in the expressions (Lee 1994). To achieve this, some extra variables, the *state variables*, are introduced. They are not measured, but can be reconstructed from the measured input-output data. This is especially useful when there are several output signals—that is, when $y(t)$ is a vector.

The state-space representation looks like

$$x(t+1) = Ax(t) + Bu(t) + Ke(t)$$
$$y(t) = Cx(t) + Du(t) + e(t)$$

Here, $x(t)$ is the vector of state variables. The model order is the dimension of this vector. The matrix K determines the disturbance properties. Notice that if $K = 0$, then the noise source $e(t)$ affects only the output, and no specific model of the noise properties is built. This corresponds to $H = 1$ in the general description above, and is usually referred to as an *output-error model*. Notice also that $D = 0$ means that there is no direct influence from $u(t)$ to $y(t)$. Thus, the effect of the input on the output all passes via $x(t)$ and will thus be delayed at least one sample. The first value of the state variable vector *x(0)* reflects the initial conditions for the system at the beginning of the data record. When dealing with models in state-space form, a typical option is whether to estimate *D*, *K*, and *x(0)* or to let them be zero.

9.4.5 How to Know Which Structure and Method to Use

There is no simple way to find out "the best model structure"; in fact, for real data, there is no such thing as a *best* structure. It is best to be generous at this point. It often takes just a few seconds to estimate a model, and by the different validation tools, one can quickly find out if the new model is any better than the ones we had before. There is often a significant amount of work behind the data collection, and spending a few extra minutes trying out several different structures is usually worthwhile.

9.5 Common Features of Commercial Identification Packages

Different MPC vendors have different features and model structure in their model identification packages. Some of those features described in Chapter 16.

Some of the common features usually available in identification software are given below:

- It provides a powerful tool to inspect/manipulate data and generate multiple input multiple output (MIMO) system models.
- All dependent and independent variables are considered simultaneously.
- It allows operator interaction during plant testing.
- Process does not need to have steady initial or terminal conditions.
- Multiple model forms and structures are accommodated.
- Data segmentation is permissible.
- Performance is given in terms of:
  - Step response curves
  - Correlation curves
  - Confidence plots
  - FIR null hypothesis test and ranking (statistics)
  - Time series prediction per CV
  - Power and uncertainty spectrum
  - Residual error and cross correlation
- System models can be automatically chosen, based on open-loop prediction performance.
- Both continuous and discrete time models are generated.
- Cross-validation analysis is easily accommodated.

References

DMC Corp. [DMC] (1994). Technology overview. Product literature from DMC Corp., July 1994.

Gustavsson I., L. Ljung and T. Söderström (1977). Identification of processes in closed loop—identifiability and accuracy aspects. *Automatica*, 13, 59–75.

Honeywell Inc. (1995). RMPCT concepts reference. Product literature from Honeywell, Inc., October 1995.

Hjalmarsson, H., M. Gevers, F. de Bruyne (1996). For model-based control design, closed-loop identification gives better performance. *Automatica*, 32(12), 1659–1673.

Larimore, W. E. (1990). Canonical variable analysis in identification, filtering and adaptive control. Proceedings of 29th IEEE CDC, Honolulu, Hawaii, pp. 596–604.

Lee, J. H., Morari, M., & Garcia, C. E. (1994). State-space interpretation of model predictive control. *Automatica*, 30(4), 707–717.

Ljung, L. (1985). Asymptotic variance expressions for identified black-box transfer function models. *IEEE Trans. Autom. Control*, Vol. AC-30, pp. 834–844.

Ljung. L. (1999). System Identification: Theory for the User. Second Edition. Englewood Cliffs, NJ: Prentice-Hall.

Snow, W. P., K. F. Emigholz and Y. C. Zhu (2001). Increase MPC Project Efficiency by using a Modern Identification Method. ERTC Computing, 18–20 June, 2001, Paris.

Van Overschee, P., and B. De Moor (1994). N4SID: subspace algorithms for the identification of combined deterministic-stochastic systems. *Automatica*, 30(1), 75–93.

Verhaegen, M. (1994). Identification of the deterministic part of MIMO state space models given in innovations form from input-output data. *Automatica*, 30(1), 61–74.

Zhu, Y. C. (2001). Multivariable System Identification for Process Control. Elsevier Science, Oxford.

Zhu, Y. C. and F. Butoyi (2002). Case studies on closed-loop identification for MPC. *Control Engineering Practice*. 10, 403–417.

10

Soft Sensors

10.1 What Is a Soft Sensor?

Chemical process industries (CPI) are usually heavily instrumented with large number of flow, temperature, pressure transmitters, and online composition analyzers. Modern chemical plants are equipped with data historian that constantly collect and store online data from all of these sensors at every seconds or minutes. Approximately two decades ago, researchers started to make use of the huge data inventory by building predictive models based on these data. These predictive models are commonly called *soft sensors*. Other common equivalent terms for predictive sensors in CPI are inferential sensors, virtual online analyzers, and observer-based sensors. By definition, an inferential or soft sensor is a mathematical relation that calculates or predicts a controlled property using other available process data. Examples of soft sensors are many:

- Polymer process: melt flow index (MFI) or melt/flow ratio (MFR), density
- Refining or petrochemical process: Product impurity ($C5^+$, $C4^-$ etc.), petroleum fraction boiling points IBP, EP, and petroleum fraction cold properties (such as flash, freeze, cloud point)

10.2 Why Soft Sensors Are Necessary

- When it is very difficult or costly to measure an important parameter online, such as distillation tower top product impurity, soft sensors are used to predict that inferential property from other easily measurable parameters such as top temperature and pressure.
- Sometimes soft sensors are used as back-ups to an existing analyzer to reduce or eliminate dead time, both from the process and the analyzer cycle.
- Soft sensors provide redundancy for a limited time in case the analyzer is unavailable because of maintenance or calibration, for example.
- The original and still most dominant application area of soft sensors is the prediction of process variables, which can be determined as low sampling rate (online analyzer) or through offline laboratory analyses only. In most cases, these variables directly or indirectly affect the process output quality. Thus, the real-time information with a high-sampling rate is very important for these variables for process control and

Multivariable Predictive Control: Applications in Industry, First Edition. Sandip Kumar Lahiri.
© 2017 John Wiley & Sons Ltd. Published 2017 by John Wiley & Sons Ltd.

management. For these reasons, it is of great interest to deliver additional information about these variables at higher sampling rates and/or lower financial burden, which is exactly the role of soft sensors.

Major areas where industrial soft sensors are proven successful are summarized in the sections that follow.

10.2.1 Process Monitoring and Process Fault Detection

Panel operators monitor thousands of process parameters. In case of process fault or sensor malfunction, it is the panel operator, based on his readiness, experience and knowledge who must identify the fault and take preventive or corrective actions to minimize the impact on production quality and quantity. The role of process monitoring soft sensors is based on the large number of historical data, to build statistical multivariate features that are relevant for the description of process state. By presenting the predictive process state or the multivariate features, the soft sensor can support the process operators and allow them to make faster, better and more objective decisions. Mostly, principal component analysis (PCA) are used to build such type of fault detection saw sensors.

10.2.2 Sensor Fault Detection and Reconstruction

As already mentioned, modern chemical plants are equipped with a large number of sensors like temperature, pressure, flow, and composition analyzers. Out of these large numbers of process parameters, some are key parameters without which process operators will find it difficult to maintain the product quality and flow rate. As these process parameters are measured by hardwire-based sensors, there is always a possibility that these physical sensors will fail. Once a faulty physical sensor is detected and identified, it can be reconstructed or the hardwired sensor can be replaced by another soft sensor, which is trained to act as a backup soft sensor of the hard-wired measuring device. If the backup soft sensor is proved effective as a replacement of the physical sensors, then it will give the operator as a standby indication of the process variables. In some occasions, the robust soft sensors can replace the measuring device even it is in working condition. The software tool can be easily maintained, is not subject to mechanical failure, is easy to implement in DCS, and can therefore provide substantial financial advantage.

10.2.3 Use of Soft Sensors in MPC Application

One of the major objectives of MPC is to push the process to its constraints to get maximum benefit. One of the major constraints of every commercial process is the product quality spec (i.e., maximum impurity limit in product). In most cases, product impurity is measured by offline laboratory analysis due to industrial regulations. In some cases, quality is measured by online analyzer, which has a large dead time. Due to this, both offline and online analyses cannot be efficiently used for optimization of process in real time. To optimize the process and to push the process to its soft and/or hard limit, MPC needs this quality spec to be made available in a real-time basis. MPC can then adjust other process parameters in such a way that these quality specs are always maintained. This prediction capability of MPC is only possible if reliable soft sensors are available for these quality parameters. This is a great application of soft sensors to build MPC models based on it. Most of the time, these quality parameters are made for CVs with tighter limits, and the job of the MPC is to adjust the relevant MVs so that these CVs are always within their acceptable limits.

10.3   Types of Soft Sensors

There are four types of soft sensors (refer Figure 10.1), as described in the following sections.

10.3.1   First Principle-Based Soft Sensors

If the phenomenology of the process is well understood and can be expressed by first principle-based material, energy and momentum balance, then these types of soft sensor can be developed.

10.3.1.1   Advantages
- These types of model-based software are best, as they have excellent extrapolation capability. They also give light on inherent physics of the process.
- As the model equations are available, it is easy to know which parameters are responsible to drive a quality parameter outside its normal limit.

10.3.1.2   Disadvantages
Soft sensors are computationally intensive for real-time application. Note that the main advantage of soft sensors is that they provide real-time information of key quality parameters, so that the information can be used to control the process. It is difficult to utilize this type of complex differential and algebraic equations in real-time application of MPC (say, 1 minute execution time, typically), which needs more time to get the solution:

Figure 10.1 Types of soft sensors

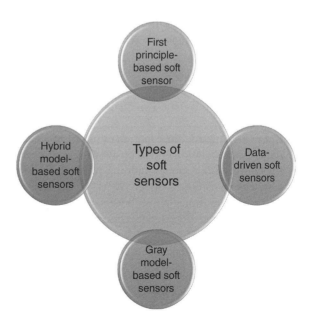

- As most of the modern industrial processes are very complex, their phenomenology (like reactor kinetics, catalyst behavior, polymer reactors, etc.) is not fully understood. So it is very difficult and time-consuming to develop a credible phenomenological model that can be implemented in commercial plans.
- Most of the phenomenological models were developed in R&D laboratory environment with a lot of assumptions. The validity of these assumptions in real-life commercial plants is questionable, and thus such soft sensor models suffer credibility issues in commercial application.

10.3.2 Data-Driven Soft Sensors

With the development of process data historian and fast computers, a large number of process data are stored every minute. This gives scope to develop data-driven soft sensors. Statistical or black-box models are developed from huge amounts of historical process data.

10.3.2.1 Advantages
- Easy to develop, where phenomenology of process is not well understood.
- Makes use of large amounts of data to derive inferential black-box model. Very accurate predictive models can be developed in commercial plants as reported in literature. These types of models can be developed quickly with the help of various model-building software like Aspen IQ, MATLAB etc.
- Lots of industrial application have already been made with many black-box modeling technique, and they have been running effectively for years as reported in literature.

Artificial neural network (ANN), principal component analysis (PCA), partial least square (PLS), and support vector regression (SVR) are some of the black-box modeling technique which are well established. They are used to build such inferential model for soft sensors. These data-driven techniques are easily available and can be utilized for a large variety of soft sensor where phenomenology of process is not well understood.

10.3.2.2 Disadvantages
- As these models are developed from process data, they are as good as the accuracy of collected data. Erroneous data can lead to inaccurate models (garbage in, garbage out).
- Data cleaning, missing data, data interpretations, process lag identification, process dynamics evaluation and so on are some common issues that need to be tackled effectively during the model-building phase. If these were not tackled effectively, robust soft sensors cannot be made.
- Extrapolation capability of these data-driven models are limited, and their credibility is doubtful once they depart from a range of data utilized to build them.

10.3.3 Gray Model-Based Soft Sensors

Most of the time, some process knowledge or phenomenology is available for industrial processes. This type of model-building technique makes use of good part of both phenomenological modeling and data-driven modeling. The developer first tries to predict the output using available phenomenological modeling. As the phenomenological model could not explain the whole process dynamics and interactions, it alone cannot accurately predict the output. A residual error will generate when compared actual output and predicted output. The residual error is then modeled by black-box type algorithms like ANN, PCA, PLS, and SVR. Output from first-principle models and output from black-box models are combined to generate the final output.

10.3.3.1 Advantages

- This type of modeling technique is more preferred than completely black-box modeling, as it somehow captures the underlying physics of the process. Also, it has more extrapolation capability than black-box models.
- It is not purely dependent on data, so it is not so error prone on the data accuracy or data quality.

10.3.4 Hybrid Model-Based Soft Sensors

Sometimes two type of black-box modeling techniques are combined to generate a final hybrid model. Hybrid modeling techniques generate predictions that are more accurate and robust than single black-box techniques (see Lahiri & Khalfe 2009; and Lahiri & Ghanta 2010). ANN plus PCA, ANN plus PLS, and ANN plus SVR techniques capture the benefits of both ANN and other techniques (see Yan, Shao, & Wang 2004; Kadlec, Gabrys, & Strandt 2009; Lahiri and Khalfe 2009; Li, Zhang, & Li 2005; and Li, Zhang, & Qian 2005).

10.3.4.1 Advantages

Each black-box modeling technique has some unique advantages and disadvantages. This type of hybrid modeling makes use of the strengths of each technique and thus generates more accurate predictions.

10.4 Soft Sensors Development Methodology

This section describes the typical steps for soft sensor development. The presented procedure is generic and can be applied for any data-driven soft sensor development.

Steps involve to develop reliable soft sensors are shown in Figure 10.2.

10.4.1 Data Collection and Data Inspection

Data-driven soft sensors are made from a large number of historical data. The quality of developed soft sensors is as good or as bad as the input data quality. The following jobs are usually performed in this step:

1) *Preparation before data collection:* Calibration of relevant transmitters and analyzers and accuracy check of collected data should be performed. This step is important to ensure accuracy of collected data.
2) *Setting of data historian:* Installation of data historian software (PIMS) and starting of data collection of all relevant tags with time stamp.

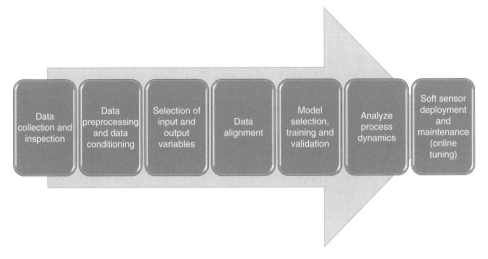

Figure 10.2 Steps involved in developing reliable soft sensors

3) *Identification of laboratory data:* Data must be identified for soft sensor and setting of laboratory information management system (LIMS) with time stamp.
4) *Inspection of data to get bird's eye view of data structure:* Inspection leads to the identification of obvious problems that may be handled at this initial stage (e.g., locked variables having constant value etc.).
5) *Assessment of the target variable:* It has to be checked to see if there is enough variation in the output variables and if this can be modeled at all.
6) *Evaluation of the steady-state part of the data:* Normally, soft sensors are built with data at steady state. Data in an unsteady state (i.e., when temperature, pressure etc. of the process is changing) might not be representative enough to build soft sensor.

10.4.2 Data Preprocessing and Data Conditioning

Data gathered from process industries normally involve the following critical characteristics:

- Data contain spikes or very high value.
- Data contain noise due to process itself or from measuring transmitters.
- Data may contain missing value or value may freeze.

The following tasks are performed in this step:

- Data cleaning
- Outlier detection and replacement
- Handling of missing data
- Selection of relevant variables (i.e., feature selection)

A systematic approach for data preprocessing and soft sensor development is given in Lin (2007).

10.4.2.1 Outlier Detection and Replacement

Outliers are commonly defined as observations that are not consistent with the majority of data, including missing data point and observation that deviate significantly from normal values. Reasons for outlier in process industry data are many (e.g., malfunction of process equipment or measuring transmitters, nonfunctioning of data collection system, etc.) can generate outlier process data. Outlier detection and removing them from the data set are critical for soft sensor development because undetected outliers reduce the performance of the final soft sensor model.

10.4.2.2 Univariate Approach to Detect Outliers

The 3σ edit rule is a popular univariate approach to detect outliers

$$\left| x(i) - \bar{x} \right| > 3\sigma$$

where $\bar{x}$ is the mean of the data sequence. This method labels outliers when data points are 3σ or more standard deviations from mean.

Median absolute deviation (MAD) is defined as

$$MAD = 1.4826 \, median \left[\left| x_i - x^* \right| \right]$$

where x^* is the median of the data sequence. The factor 1.4826 is chosen such that expected MAD is the standard deviation for normally distributed data (Davies & Gather, 1981). With the limit of $x_{med} \pm 3x_{MAD}$, the Hampel identifier identifies most outliers successfully.

10.4.2.3 Multivariate Approach to Detect Outliers (Lin 2007)

1) Principal component analysis (PCA) is a multivariate statistical method that reduces the data dimensionality by projecting the data matrix to a lower dimensional space with the help of the loading vectors. The loading vectors corresponding to the k largest eigenvalues can capture the variations of the data and thus contains most of the information. The fitness between data and the model can be calculated using the residual matrix and Q statistics that measure the distance of a sample from the space of the PCA model (Jackson & Mudholkar, 1979). Hotelling's $T2$ statistics indicate the distance between a particular data from the multivariate mean of the data; thus, these statistics provide an indication of variability within the normal subspace (Wise, 1991).

2) The combined Q and $T2$ tests are used to detect outliers. Given the significance level for the Q and $T2$ statistics, measurements with Q or $T2$ values over the threshold are classified as outliers. Outliers are located outside of the 99 percent confidence ellipse. In another approach, data outliers are detected in two phases. In the first phase, it removes outliers using a univariate approach, then carries out a classic PCA on the new data set. In the second phase, it applies the multivariate approach and calculates covariance matrix. The proposed procedure uses the ellipsoidal multivariate trimming (MVT) approach (Devlin, Gnanadesikan, & Kettenring, 1981). This trimming method iteratively detects bad data based on the squared *Mahalanobis* distance:

$$d_i^2 = \left(x_i - x_i^* \right)^T S^{*-1} \left(x_i - x_i^* \right)$$

where x^* is the current robust estimation of the location and S^* is the robust estimation of the covariance matrix. Since the data set has been preprocessed with a *Hampel identifier*, 95 percent of data with smallest *Mahalanobis* distance are retained in the next iteration.

Devlin et al. (1981) suggest that the iteration proceeds until the average absolute change in Fisher z transforms of the elements of the correlation matrix between two successive iterations is less than a predefined threshold, or the maximum number of iteration is reached.

10.4.2.4 Handling of Missing Data

Missing data are single samples or consequent sets of samples, where one or more variables (i.e., measurements) have a value that does not reflect the real state of the physical measured quantity. The affected variables usually have values like $\pm \infty$, 0, or any other constant value.

Missing values in the context of process industry have various causes given below:

- Sensor hardware may fail.
- Data communication between sensors and DCS may fail.
- Data historian may be stopped from collecting data.

Since most of the techniques applied to data-driven soft sensing cannot deal with missing data, a strategy for their replacement must be implemented. There are different strategies to replace missing values:

- Replace missing value with average value of the affected variable (not recommended).
- Delete the whole record, which consists of missing value of some variable such as case deletion (Scheffer 2002).
- Replace the missing value with maximum-likelihood multivariate statistics calculations. (see, e.g., Walczak & Massart 2001b) (efficient approach).

Different researchers propose different techniques to replace missing values. As the data set from process industry is very large, these techniques must be automated, as manual replacement is cumbersome and impossible.

Scheffer (2002) discusses following two different approaches to replace missing value: (i) single imputation where the missing values are replaced in a single step (using e.g. mean/median values) and (ii) multiple imputation, which are iterative techniques where several imputation steps are performed.

Schafer and Graham (2002) propose maximum-likelihood and Bayesian multiple imputation techniques to handle missing data.

In Chen and Chen (2000), an iterative reweighed least squares technique is applied to calculate missing data values. Walczak and Massart (2001a, 2001b) proposed both PCR/PLS-based techniques and a maximum-likelihood-based algorithm for dealing with missing data. An alternative approach to dealing with missing data in a probabilistic framework was published in Gabrys (2002). This work particularly focuses on missing data treatment in the context of decision making and diagnostic analysis.

10.4.3 Selection of Relevant Input Output Variables

Input selection is a key step to model input–output relationships during soft sensor model building. Here are some guidelines:

1) Simple model is the best. Minimum number of input for modeling of output is best. More the input, more will be the noise in the predicted output.
2) Based on process knowledge, find out probable inputs that have a cause-and-effect relationship with the output.
3) Try to select intensive property input like temperature, pressure etc. rather than flow. For example, to model an impurity at distillation column overhead product, the obvious input would be tower top temperature and pressure, delta P of column, reflux ratio, etc. It is advisable to avoid taking input as reflux flow or overhead product flow.
4) Input should be independent of each other (i.e., there should not be any data colinearity between the input). As, for example, select only one representative tray temperature for a distillation column rather than choosing three or four tray temperatures that are dependent on each other.
   First, start with a minimum number of inputs and then increase one by one until there is significant performance improvement of the developed model. If prediction performance is not improved significantly by adding one extra input, don't add that input.

10.4.4 Data Alignment

Soft sensing is often applied in multivariate systems with several operating sample rates. In many industrial chemical processes, normally product quality parameters are measured offline in a laboratory (once in every 8 hours or so) or by online analyzer with long dead time (typically 15 minutes). The input variables like temperature and pressure are measured and recorded every few seconds or minutes. So it is necessary to align the data in proper time scale. It is absolutely necessary for laboratory data to be time stamped properly and align with other continuous data in proper time scale.

10.4.5 Model Selection, Training, and Validation (Kadlec 2009; Lin 2007)

This is the most critical step in soft sensor building. As a model is the heart of soft sensor, proper and accurate model selection is key to its performance. There are a lot of options available for model selection, but no clear-cut guidelines are available to select which model at what conditions. In most cases, the developer selects the type of model based on personal choice and expertise. This can be detrimental to developing a good model. The best approach is to remain open minded for all the model types. A good practice is to start with a simple model type or structure (e.g., linear regression model) and gradually increase model complexity as long as significant improvement in the model's performance can be observed (using, e.g., the Student's *t*-test, Gosset 1908).

During the model-building phase, performance of individual model can be judged by unseen validation data (Hastie et al. 2001; Weiss & Kulikowski 1991). The same approach can also be applied to the parameters selection of the preprocessing methods, like, for instance, variable selection. Normally, data-driven models need large amounts of data that are usually available in modern industry. However, in some instances where lab data are used, very small amounts of data might be available. Additionally, for some industrial processes where there are few reliable lab data, statistical error-estimation techniques like *K*-fold cross-validation (Kohavi 1995) can be applied. This method makes optimal use of the available data by partitioning it in such a way that all of the samples are used for the model performance validation. Another alternative in these circumstances is to apply statistical resampling methods, as, for example, bagging (Breiman 1996) and boosting (Freund & Schapire 1997).

In the case of the first method, a set of training data sets is generated by randomly drawing samples (with replacement) from the available data and training one model for each of the random sets. The final model is obtained by averaging over the particular models' predictions. In contrast to this, in the case of boosting, the probability of each sample to be drawn is not random but related to the prediction error of the model given in the data sample. Additionally, in the case of boosting, the weights of the contributions of the particular models are calculated based on the model's performance on a validation data set.

After finding the optimal model structure and training the model, the trained soft sensor performance has to judge the new validation data set once again (Weiss & Kulikowski 1991). Mean squared error (MSE), which measures the average square distance between the predicted and the correct value, is the most popular performance evaluation techniques for soft sensors. Another way of performance judgment is using visual representation of the predictions. In these, the four-plot analysis is a useful tool since it provides useful information about the relation between the predictions and the correct values, together with the analysis of the prediction residuals (Fortuna 2007). A disadvantage of the visual methods is that they require assistance of the model developer, and the final decision if the model performs adequately, is up to the subjective judgment of the model developer.

To evaluate whether the developed model resembles the underlying physics of the process, Fortuna (2007) stresses the necessity for the application of process knowledge during the soft sensor development phase.

10.4.6   Analyze Process Dynamics

The dead time refers to the time lag that occurs between when an independent variable value changes and when the dependent variable starts to change in response. This step is used only for dynamic inferential and provides a very coarse estimate of the dead times. Normally, a genetic algorithm or differential evolution is used to find out dead times of each individual variable:

- In the analyze dynamics step, dead times are recomputed, and dynamic filters are computed based on analyzing model training data in terms of process movement toward steady state.
- As we stated earlier, an independent variable's dead time represents the time delay between when its value changes and when the model's dependent variable value starts to change.
- The variable's dynamic filter then represents the time delay between the start of the independent variable's change and when the dependent variable reaches 63.2 percent of its final steady-state value.

- Like the align data step, the analyze dynamics step only applies to dynamic inferential. It is not available for steady state inferential.
- The dead times and dynamic filters are computed for only those independent variables that were selected for the final model in the select variables step. This is a change from the align data step, in which dead time estimates were computed for all independent variables.
- Use genetic algorithm for analyze dynamics step.

Genetic algorithm assumes the dead time randomly based on process knowledge and then stochastically continually changes it until the best matching performance with validation data set is achieved. It finds the "best" combination of dead times and filter times for input variables using R^2 statistics to detect which one is the best.

10.4.7 Deployment and Maintenance

After developing and deploying the soft sensor, it must be maintained and tuned on a regular basis. It is common in industry that the performance of soft sensor deteriorates over time. Reasons are many. The underlying process may change (e.g., catalyst selectivity, yield may change over time, heat exchanger may foul, etc.), Measuring transmitters data may drift, and analyzer reading may change due to recalibration etc. All of these can cause the performance of the soft sensors to deteriorate and must be compensated for by adapting or redeveloping the model.

Currently, most of the soft sensors do not provide any automated mechanisms for their maintenance. In some cases, soft sensors software comes with a MPC software package that has features for automatic lab data entry and auto bias update.

Industrial software for soft sensors usually contain two modules: Offline module is used to build the accurate soft sensor model and online module is used to automatically update the bias based on lab analysis. Various features of these types of soft sensing software are described in Chapter 16. Normally, in industry, performance of soft sensors is evaluated by panel operators or production engineers based on their perception of how close the soft sensors value matches with the lab data. In many occasions, it has been observed the soft sensors performance deteriorates over time. In literature, researchers tried various adaptive approaches to update the soft sensors model based on its performance. The majority of these approaches are based on adaptive versions of the PCA or PLS, like Moving Window PCA (Wang et al. 2005) or the Recursive PCA (Li, Zhang, & Qian 2005). All of these methods rely on periodical or continuous adaptation of the principle component base. In neuro-fuzzy-based soft sensors such as Macias and Zhou (2006); Atkeson, Moore, & Schaal, (1997); Kadlec and Gabrys (2008a), automatic updating are based on the deployment of new units in the neural structure of the model once a new state of the data is found.

Most of these auto model update methods are still limited to research publications, and very few are really applied in actual industry.

Despite the methods for the automated soft sensor adaptation, the model operators still play an important role, as it is their judgment and knowledge of the underlying process that decides the way the parameters of the individual adaptation methods are selected (e.g., the length of the window in case of the moving window technique, or a threshold for the deployment of a new receptive field in case of the neuro-fuzzy methods).

10.5   Data-Driven Methods for Soft Sensing

This section outlines and provides further references to the most popular techniques for soft sensor development as employed in process industries. These are the multivariate principle component analysis, partial least squares, artificial neural networks, neuro-fuzzy systems and support vector machines. As each method requires a lot of discussion, it is kept outside of the scope of this book. However, a preliminary introduction is given here for each method.

10.5.1   Principle Component Analysis

The PCA algorithm reduces the number of variables by building linear combinations of them. This is done in such a way that these combinations cover the highest possible variance in the input space and are additionally orthogonal to each other. In the context of the process industry data this is a very useful feature because the data there are often colinear. In this way the collinearity can be handled and the dimensionality of the input space can be decreased at the same time. The PCA is usually applied as pre-processing step followed by the actual computational learning method.

10.5.1.1   The Basics of PCA

Principal component analysis (PCA) is a data transformation method that rotates data such that the principal axis of the data is in the direction of maximum variation. We can view the rotated data on the new principal axes. The coordinates of the data in this new coordinate system are known as principal component scores. These are essentially projections of the data onto the principal axes. The principal axes (components) are essentially *vectors* in the original variable space. These vectors are known as principal component loadings.

10.5.1.2   Why Do We Need to Rotate the Data?

Although data sets may contain many variables (i.e., n dimensions), variation is often limited to a few key directions. By reorienting our coordinate system to align with these key (multivariable) directions we can often find that some of the directions perpendicular to this space contain very limited variation and can therefore be ignored. Hence, by reorienting the way we view the data, we can squeeze as much information as possible into the three key principal directions.

10.5.1.3   How Do We Generate Principal Components?

The principal components represent the selection of a new coordination system obtained by rotating the original variables and projecting them into the reduced space defined by the first few principal components, where the data are described adequately and in a simpler and more meaningful way.

The principal components are ordered such that the first one describes the largest amount of variation in the data, the second one the second largest amount of variation, and so on. With highly correlated variables, one usually finds that only a few principal components are needed to explain most of the significant variation in the data.

10.5.1.4   Steps to Calculating Principal Components

It is shown in statistics that for a normalized data matrix Xs (i.e., $Xs^T Xs = 1$), then the eigenvector of Xs^TXs (the covariance matrix) with largest eigenvalue will point in the direction of maximum variance in the data. In fact, the variance of the data along each eigenvector is equal to the eigenvalue of that eigenvector. The proof also shows that eigenvectors are, by definition, orthogonal to each other. Hence, the eigenvectors of the covariance matrix are the principal components of the data.

- Consider a normalized data matrix X (each column has zero mean and unit standard deviation) of size n rows and m columns $(n \times m)$
  The covariance of $X = X^TX$
- Calculate all eigenvectors and eigenvalues of X^TX
- Place the eigenvectors in order of decreasing eigenvalue. This matrix is called the *Loadings Matrix P* $(m \times m)$
- To calculate the principal component scores (i.e., the projections of original normalized data onto this new eigenvector basis), multiply the normalized matrix X by the loadings matrix P.

Scores matrix will be: $\mathbf{T} = \mathbf{XP}$ $(n \times m)$

The loading vectors P are orthonormal and provide the directions with maximum variability. The T scores from the different principal components are the coordinates for the objects in the reduced space. They are uncorrelated and therefore are measuring different underlying latent structures in the data. By plotting the scores of one principal component vs. another, one can easily see which of the objects have similarities in their measurements and form clusters, and which are isolated from the others and therefore are unusual objects or outliers. The power of PCA arises from the fact that it provides a simpler and more parsimonious description of the data covariance structure than the original data.

Although the PCA is a well-established and powerful algorithm, it has several drawbacks and limitations. One of the limitations is that the pure PCA can only effectively handle linear relationships (correlations) of the data and thus cannot deal with nonlinearity of the data. This limitation has been solved by extending the original PCA algorithm, as discussed in the previous paragraph.

Another issue is the selection of optimal number of principal components. This problem is most commonly approached by using cross-validation techniques.

Another problem is that the principal components describe very well the input space but do not reflect the relation between the input and the output data space, which actually has to be modeled. A solution to this problem is given by the partial least squares method discussed in the next section.

10.5.2   Partial Least Squares

This algorithm, instead of focusing on covering the input space variance, pays attention to the covariance matrix that brings together the input and the output data space. The algorithm decomposes the input and output space simultaneously while keeping the orthogonality constraint. In this way, it is assured that the model focuses on the relation between the input and output variables.

A general description of the PLS technique is provided in Geladi and Esbensen (1991) and Abdi (2003). As PLS is a very popular technique in chemical engineering and in chemometrics, there are several publications dealing with the application aspects of PLS to this domain (Frank & Friedman 1993; Kourti 2002).

10.5.3 Artificial Neural Networks

In the last decade, artificial neural networks (ANNs) have emerged as attractive tools for nonlinear process modeling, especially in situations where the development of phenomenological or conventional regression models becomes impractical or cumbersome. ANN is a computer modeling approach that learns from examples through iterations without requiring a prior knowledge of the relationships of process parameters and is, consequently, capable of adapting to a changing environment. It is also capable of dealing with uncertainties, noisy data, and nonlinear relationships.

ANN modeling has been known as *effortless computation* and readily used extensively because of its model-free approximation capabilities of complex decision-making processes (Lahiri 2010). The advantages of an ANN-based model are the following:

1) It can be constructed solely from the historic process input-output data (example set).
2) Detailed knowledge of the process phenomenology is unnecessary for model development.
3) A properly trained model possesses excellent generalization ability owing to which it can accurately predict outputs for a new input data set.
4) Even multiple input-multiple output (MIMO) nonlinear relationships can be approximated simultaneously and easily.

Owing to their several attractive characteristics, ANNs have been widely used in chemical engineering applications such as steady-state and dynamic process modeling, soft sensor building, process identification, yield maximization, nonlinear control, and fault detection and diagnosis (Lahiri 2010).

The most widely utilized ANN paradigm is the multilayered perceptron (MLP) that approximates nonlinear relationships existing between an input set of data (causal process variables) and the corresponding output (dependent variables) data set. A three-layered MLP with a single intermediate (hidden) layer housing a sufficiently large number of nodes (also termed neurons or processing elements) can approximate (map) any nonlinear computable function to an arbitrary degree of accuracy. It learns the approximation through a numerical procedure called *network training*, wherein network parameters (weights) are adjusted iteratively such that the network, in response to the input patterns in an example set, accurately produces the corresponding outputs. There exists a number of algorithms—each possessing certain positive characteristics—to train an MLP network, for example, the most popular error-back-propagation (EBP), Quickprop and Resilient Back-propagation (RPROP). Training of an ANN involves minimizing a nonlinear error function (e.g., root-mean-squared-error, RMSE) that may possess several local minima. Thus, it becomes necessary to employ a heuristic procedure involving multiple training runs to obtain an optimal ANN model whose parameters (weights) correspond to the global or the deepest local minimum of the error function. The building of a back-propagation network involved the specification of the number of hidden layers and the number of neurons in each hidden layer. In addition, several parameters, including the learning rule, the transfer function, the learning coefficient ratio, the random number seed, the error minimization algorithm, and the number of learning cycles had to be specified.

10.5.3.1 Network Architecture

The MLP network used in the model development is depicted in Figure 10.3. As shown, the network usually consists of three layers of nodes. The layers described as input, hidden, and output layers comprise N, L, and K number of processing nodes, respectively. Each node in the input (hidden) layer is linked to all the nodes in the hidden (output) layer using weighted connections. In addition to the N and L number of input and hidden nodes, the MLP architecture also houses a bias node (with fixed output of +1) in its input and hidden layers; the bias nodes are also connected to all the nodes in the subsequent layer, and they provide additional adjustable parameters (weights) for the model fitting. The number of nodes (N) in the MLP network's input layer is equal to the number of inputs in the process whereas the number of output nodes (K) equals the number of process outputs. However, the number of hidden nodes (L) is an adjustable parameter whose magnitude is determined by issues, such as the desired approximation and generalization capabilities of the network model.

10.5.3.2 Back Propagation Algorithm (BPA)

The back propagation algorithm (BPA) modifies network weights to minimize the MSE between the desired and the actual outputs of the network. Back propagation uses supervised learning in which the network is trained in using data for which inputs as well as desired outputs are known. Once trained, the network weights are frozen and can be used to compute output values for new input samples. The feedforward process involves presenting an input data to input layer neurons that pass the input values onto the first hidden layer. Each of the hidden layer nodes computes a weighted sum of its input and passes the sum through its activation function and presents the result to the output layer. The goal is to find a set of weights that minimize mean squared error. A typical back propagation algorithm can be given as follows.

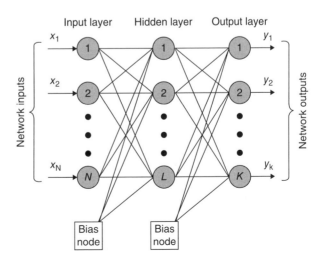

Figure 10.3 Artificial neural network architecture

The MLP network is a nonlinear function-mapping device that determines the K dimensional nonlinear function vector f where $f: X \to Y$. Here, X is a set of N-dimensional input vectors ($X = \{xp\}$; $p = 1, 2,..., P$ and $x = [x_1, x_2,..., x_n,..., x_N]T$), and Y is the set of corresponding K-dimensional output vectors ($Y = \{yp\}$; $p = 1, 2,..., P$ where $y = [y1, y2,..., yk,..., yK]T$). The precise form of f is determined by (1) network topology, (2) choice of the activation function used for computing outputs of the hidden and output nodes, and (3) network weight matrices W_H and W_O (they refer to the weights between input and hidden nodes, and hidden and output nodes, respectively). Thus, the nonlinear mapping can be expressed as

$$y = y(x;W)$$

where $W = \{W_H, W_O\}$. This equation suggests that y is a function of x, which is parameterized by W. It is now possible to write the closed-form expression of the input-output relationship approximated by the three-layered MLP as

$$y_k = f_2\left[\sum_{l=0}^{L} w_{lk}^0 f_1\left[\sum_{n=0}^{N} w_{nl}^H x_n\right]\right] k = 1,2,.......,K$$

where y_k refers to the kth network output; f_2, f_1 denote the nonlinear activation functions; w_{lk}^0 refers to the weight between lth hidden node and kth output node; w_{nl}^H is the weight between nth input and lth hidden node; and x_n represents the nth network input.

Note that in the above equation, the bias node is indexed as the 0th node in the respective layer. In order that an MLP network approximates the nonlinear relationship existing between the process inputs and the outputs, it needs to be trained in a manner such that a prespecified error function is minimized. In essence, the MLP training procedure aims at obtaining an optimal set (W) of the network weight matrices W_H and W_O, which minimize an error function.

The application aspects of large number of ANN variants, including dynamic and adaptive ANN, are discussed in Principe et al. (2000). This book is especially recommended for soft sensor modeling, as there is a large number of mutual topics between the book and the process industry applications of ANN. Detailed discussion of some problems of process industry data and of the suitability of ANN to solve this problem is provided in Qin (1997). Apart from the discussion of the ANN issues in the process industry, the work proposes some possible solution to them.

10.5.4 Neuro-Fuzzy Systems

Neuro-fuzzy system (NFS) is a hybrid intelligent model that combines the excellent predicting capabilities of the ANN with the human-like reasoning of the fuzzy inference system (FIS) (Zadeh, 1996). It is a realization of the fuzzy system by a connectionist structure of an ANN. The amalgamation of two methods generates a learning system that provides the advantages of both of the involved techniques while at the same time dealing with their drawbacks. Another appealing property for the process industry application of the NFS models is that the technique is based on receptive fields and thus intrinsically provides means for the building of local models. An introduction to NFS is provided in Jang et al. (1997) and Nauck, Klawonn, Kruse, and Klawonn (1997). The evolving variants of NFS are very well suited to dealing with dynamic environment. These systems are called *evolving* because they adapt automatically together with the changing environment represented by the data. An evolving system is thought to be able to change its structure, to grow and shrink and to update its parameters (Angelov & Kasabov 2005). In this way, the model is able to deploy new local models related to new states of the input data if necessary.

10.5.5 Support Vector Machines

Due to their theoretical background in the statistical learning theory SVMs gained attention in the computational learning community. Their derivation and theoretical justification can be found in Vapnik (1998).

10.5.5.1 Support Vector Regression–Based Modeling

Industrial data contain noise. Normally different transmitter, signal transmissions and so on add these noises with process parameters. Normal regression techniques try to reduce the prediction error on noisy training data. This empirical risk minimization (ERM) principle is generally employed in the classical methods such as the least square methods, the maximum likelihood methods, and traditional ANN (Lahiri 2009). The formulation of SVR embodies the structural risk minimization (SRM) principle, which has been shown to be superior, to traditional empirical risk minimization (ERM) principle, employed by conventional neural networks. SRM minimizes an upper bound on the expected risk, as opposed to ERM that minimizes the error on the training data. It is this difference that equips SVM with a greater ability to generalize. The SVR algorithm attempts to position a tube around the data, as shown in Figure 10.3. ε is a precision parameter representing the radius of the tube located around the regression function (see Figure 10.4); the region enclosed by the tube is known as *e-intensive zone.*

The diameter of the tube should ideally be the amount of noise in the data. The optimization criterion in SVR penalizes those data points the y values of which lie more than ε distance away from the fitted function, $f(x)$. There are two basic aims in SVR. The first is to find a function $f(x)$ that has at most ε deviation from each of the targets of the training inputs. For the linear case, f is given by:

$$f(x) = \langle w.x \rangle + b$$

where $< a.b >$ is the dot product between a and b.

At the same time, we would like this function to be as flat as possible. Flatness in this sense means a small w. This second aim is not as immediately intuitive as the first, but still important in the formulation of the optimization problem used to construct the SVR approximation.

A key assumption in this formulation is that there is a function $f(x)$ that can approximate all input pairs (x_i, y_i) with ε precession; however, this may not be the case or perhaps some error allowance is desired. Thus the slack variable, ξi and ξi^*, can be incorporated into the optimization problem to yield the following formulation:

$$minimize \; \frac{1}{2}|w|_2 + C\sum_{i=1}^{l}\left(\xi_i + \xi_i^*\right)$$

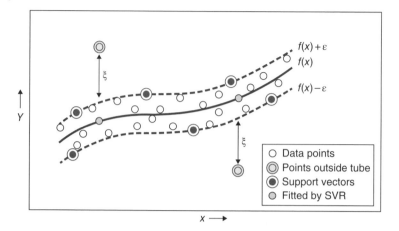

Figure 10.4 Schematic of SVR using an e-insensitive loss function

$$subject\ to \begin{cases} y_i - (w.x_i) - b \leq \varepsilon + \xi_i \\ -y_i + (w.x_i) + b \leq \varepsilon + \xi_i^* \\ \xi_i, \xi_i^* \geq 0 \end{cases}$$

In Figure 10.4, the sizes of the stated excess positive and negative deviations are depicted by ξ_i and ξ_i^*, which are termed *slack* variables. Outside the $[-\varepsilon, \varepsilon]$ region, the slack variables assume nonzero values.

The SVR fits $f(x)$ to the data in such a manner that: (i) the training error is minimized by minimizing ξ_i and ξ_i^* and (ii) w_2 is minimized to increase the flatness of $f(x)$ or to penalize over complexity of the fitting function. The constant $C > 0$ determines the trade-off between flatness (small w) and the degree to which deviation larger than ξ are tolerated, and 1 is the number of samples. This is referred to as the ε insensitive loss function proposed by Vapnik (1998), which enables as a sparse set of support vectors to be obtained for the regression.

Refer to SVM application in soft sensors in Lahiri (2009); Yan et al.; (2004), Feng et al. (2003); and Li, Zhang, and Li (2005).

10.6 Open Issues and Future Steps of Soft Sensor Development

There are several issues in the soft sensor development and maintenance.

10.6.1 Large Effort Required for Preprocessing of Industrial Data

At the development phase, a lot of effort is spent on the manual pre-processing of the data, as well as on the model selection and validation steps. To be able to deal with issues like missing values and data outliers, the model developer has to manually try different preprocessing approaches and select the one giving the best performance as estimated on the training or validation data.

10.6.2   Which Modeling Method to Choose?

Another issue is which modeling approach to choose. There are no clear-cut guidelines for this in literature. Best approach of the mentioned problem is to equip the model with the ability to select the most appropriate approach from a pool of available methods. This, of course, increases the complexity of the model, but at the same time, if implemented effectively, moves at least a part of the manual development burden to the model.

10.6.3   Agreement of the Developed Model with Physics of the Process

It is very important to understand that developed model equations of soft sensors should have a physical meaning and represent the process knowledge. In other words, the developed model should not be a sophisticated data-fitting regressed model, but also should be in agreement with the inside physics of the process. Hence, it is necessary to incorporate as much process knowledge as is available during model building phase. Remember that the first principle model is the best model. Data-driven models are tried as a compromise because first-principle models could not be made for complex industrial process. However, effort should be made to include all the available knowledge of the process in model equations so that the final model is not a pure black-box model but a grey-box model with some reflection of process knowledge.

10.6.4   Performance Deterioration of Developed Soft Sensor Model

Another major issue is related to the model maintenance. After successful launch of the soft sensor, in most cases, one can observe a gradual deterioration of the performance of the soft sensor. The decrease of the prediction quality is caused by the gradual changes in the process. Usually, after some time the performance of the model reaches an unacceptable level and the model must be retrained, or in the worst case, rebuilt from scratch.

References

Angelov, P., & Kasabov, N. (2005). Evolving computational intelligence systems. In IEEE workshop on genetic and fuzzy systems GFS2005 Grenada, Spain.

Atkeson, C. G., Moore, A.W., & Schaal, S. (1997). Locally weighted learning. *Artificial Intelligence Review*, 11(1), 11–73.

Breiman, L. (1996). Bagging predictors. *Machine Learning*, 24(2), 123–140.

Chen, J. M., & Chen, B. S. (2000). System parameter estimation with input/output noisy data and missing measurements. IEEE Transactions on Signal Processing [see also IEEE Transactions on Acoustics, *Speech, and Signal Processing*, 48(6), 1548–1558.

Davies, L., & Gather, U. (1981). The identification of multiple outliers. *Journal of the American Statistical Association*, 88, 782–801.

Devlin, S. J., Gnanadesikan, R., & Kettenring, J. R. (1981). Robust estimation of dispersion matrices and principal components. *Journal of the American Statistical Association*, 76, 354–362.

Feng, R., Shen, W., & Shao, H. (2003). A soft sensor modeling approach using support vector machines. In *Proceedings of the 2003 American control conference, 2003* (p.5).

Fortuna, L. (2007). Soft sensors for monitoring and control of industrial processes. New York: Springer.

Freund, Y., & Schapire, R. E. (1997). A decision-theoretic generalization of on-line learning and an application to boosting. *Journal of Computer and System Sciences*, 55(1), 119–139.

Gabrys, B. (2002). Neuro-fuzzy approach to processing inputs with missing values in pattern recognition problems. *International Journal of Approximate Reasoning*, 30(3), 149–179.

Gosset, W. S. (1908). The probable error of a mean. *Biometrika*, 6(1), 1–25.

Hastie, T., Tibshirani, R., & Friedman, J. (2001). The elements of statistical learning: Data mining, inference, and prediction. New York: Springer.

Jackson, J. E., & Mudholkar, G. S. (1979). Control procedures for residuals associated with principal component analysis. *Technometrics*, 21(3), 341–349.

Jang, J. S. R., Sun, C. T., & Mizutani, E. (1997). Neuro-fuzzy and soft computing. Upper Saddle River, NJ: Prentice Hall.

Kadlec, P., & Gabrys, B. (2008a). Adaptive local learning soft sensor for inferential control support. In Proceedings of the CIMCA 2008. Vienna: IEEE.

Kadlec, P., Gabrys, B., & Strandt, S. (2009). Data-driven soft sensors in the process industry. *Computers & Chemical Engineering*, 33(4), 795–814.

Kohavi, R. (1995). A study of cross-validation and bootstrap for accuracy estimation and model selection. In Proceedings of the fourteenth international joint conference on artificial intelligence, Vol. 2 (pp. 1137–1145).

Lahiri, S. K., and Khalfe, N. M. (2009). Soft sensor development and optimization of the commercial petrochemical plant integrating support vector regression and genetic algorithm. *Chemical Industry & Chemical Engineering Quarterly*, 15(3).

Lahiri, S. K., & Ghanta, K. C. (2010). Artificial neural network model with parameter tuning assisted by genetic algorithm technique: study of critical velocity of slurry flow in pipeline. *Asia-Pacific Journal of Chemical Engineering*, 5(5), 763–777.

Li, M., Zhang, Z., & Li, W. (2005). Study on least squares support vector machines algorithm and its application. In *Proceedings of the 17th IEEE international conference on tools with artificial intelligence* (pp. 1082–3409).

Li, S. J., Zhang, X. J., & Qian, F. (2005). Soft sensing modeling via artificial neural network based on PSO-Alopex. In Proceedings of 2005 international conference on machine learning and cybernetics, 2005 (p. 7).

Lin, B., Recke, B., Knudsen, J. K., & Jørgensen, S. B. (2007). A systematic approach for soft sensor development. *Computers & Chemical Engineering*, 31(5), 419–425.

Macias, J. J., & Zhou, P. X. (2006). A method for predicting quality of the crude oil distillation. In Proceedings of the 2006 international symposium on evolving fuzzy systems (pp. 214–220).

Nauck, D., Klawonn, F., Kruse, R., & Klawonn, F. (1997). Foundations of neuro-fuzzy systems. New York: John Wiley & Sons, Inc.

Principe, J. C., Euliano, N. R., & Lefebvre, W. C. (2000). Neural and adaptive systems. New York: John Wiley & Sons.

Qin, S. J. (1997). Neural networks for intelligent sensors and control—Practical issues and some solutions. *Neural Systems for Control*, 213–234.

Schafer, J. L., & Graham, J. W. (2002). Missing data: Our view of the state of the art. *Psychological Methods*, 7(2), 147–177.

Scheffer, J. (2002). Dealing with missing data. *Research Letters in the Information and Mathematical Sciences*, 3(1), 153–160.

Vapnik, V. N. (1998). Statistical learning theory. New York: John Wiley & Sons.

Walczak, B., & Massart, D. L. (2001a). Dealing with missing data. Part I. *Chemometrics and Intelligent Laboratory Systems*, 58(1), 15–27.

Walczak, B., & Massart, D. L. (2001b). Dealing with missing data. Part II. *Chemometrics and Intelligent Laboratory Systems*, 58(1), 29–42.

Wang, X., Kruger, U., & Irwin, G. W. (2005). Process monitoring approach using fast moving window PCA. *Industrial & Engineering Chemistry Research*, 44(15), 5691–5702.

Weiss, S., & Kulikowski, C. (1991). Computer systems that learn: Classification and prediction methods from statistics, neural nets, machine learning, and expert systems. San Francisco, CA: Morgan Kaufmann Publishers Inc.

Wise, B. M. (1991). Adapting multivariate analysis for monitoring and modeling dynamic systems. Ph.D. thesis. University of Washington.

Yan, W., Shao, H., & Wang, X. (2004). Soft sensing modeling based on support vector machine and Bayesian model selection. *Computers and Chemical Engineering*, 28(8), 1489–1498.

Zadeh, L. A. (1996). Fuzzy sets. World Scientific Series in Advances in Fuzzy Systems, 19–34.

11

Offline Simulation

11.1 What Is Offline Simulation?

Before commissioning the controller in an actual process plant, it is extremely important to know how the controller performs in real time but in offline mode. Offline simulation refers to running the controller in a separate offline PC to see the manipulated variable (MV)–control variable (CV) dynamic responses of the process.

11.2 Purpose of Offline Simulation

There are several purposes for offline simulations:

- It gives MPC engineers and plant process engineers an opportunity to assess the real-time performance of the developed MPC application. Normally, simulation should be run in front of the operators, operation engineers, process engineers, and control engineers of the plant, so that everybody can view the MPC dynamic performance from their own viewpoint. Buy-in from operation department is absolutely necessary. An effective, MPC application should satisfy all the stakeholders.
- In offline simulations, different high–low limits of MV and CV are set with consultation with operation department.
- Different tuning parameters of the controller are also set to make it effective on its real-time performance.
- Offline tuning of these parameters was done after observing the offline simulation responses of the controller. If controller dynamic performance is not satisfactory, if MV movements are large, if different MV does not move as per their priority, if CV errors are not acceptable for important CV as per their priority, and so on, it is time to change the different tuning parameters of the MPC controller and observe its impact until a satisfactory dynamic performance is achieved.

The main purpose of the simulation is to observe the following in an offline environment:

- Whether multivariable controller is able to maintain and control the process within reasonable time and accuracy.
- Whether correct MVs are moved to control relevant CVs. Whether MVs move as per their priority.

Multivariable Predictive Control: Applications in Industry, First Edition. Sandip Kumar Lahiri.
© 2017 John Wiley & Sons Ltd. Published 2017 by John Wiley & Sons Ltd.

- Whether MV movements are in controlled way or it gets bumpy and aggressive. Simulation is used to tune move suppression.
- Whether the optimizer performs its duty and is able to drive the process in a more economical operating zone.
- Whether the optimizer is able to push the process at its constraints or limit.
- How insufficient degrees of freedom is handled (i.e., how controller gives up some of the CVs as per their priority and able to control the process in a feasible region).

11.3   Main Task of Offline Simulation

Offline simulation has three main tasks:

1) *Setting up the simulator:* Normally, MPC software have built-in simulator facility. This task involves activating the offline simulator, setting up different tuning parameter values, making arrangement of different trends and windows of MV, CV, and DVs for visualization, and putting realistically different limits on MV and CVs so that an effective simulation study can be done.
2) *Preparation of different cases for simulator:* This task involves planning and setting up different tasks and simulation cases, which will be used for offline simulations. Note that the main purpose of this type of simulations is to judge the controller performance from control and optimization point of view. The purpose of this step is to develop and run different cases so that each and every feature of MPC application can be seen and judged. The target is to tune the offline controller so well that transfer of the offline controller to online application becomes bumpless and effortless.
3) *Offline tuning:* The purpose is to perform offline tuning and other corrections as much as possible so that the application runs effortlessly in the actual plant at real time. Offline tuning involves setting proper priorities for CVs and MVs, CV give-up priority in case of infeasible solution, and setting up optimizer speed and different co-efficient of LP and QP objective function. Changes in different tuning parameters are done on trial and error basis until a satisfactory dynamic performance of MPC controller is achieved.

11.4   Understanding Different Tuning Parameters of Offline Simulations

Before starting offline simulations, it is important to understand the concept of different tuning parameters available. It is crucial to understand how MPC works in a dynamic environment and how different tuning parameters can impact its performance. Figure 11.1 shows some basic tuning parameters commonly employed by commercial MPC software.

As discussed earlier, a working principle of MPC can be explained by three modules—namely, prediction module, steady-state optimization module, and dynamic control module. Refer to Figure 2.2 of Chapter 2 to follow how these modules work.

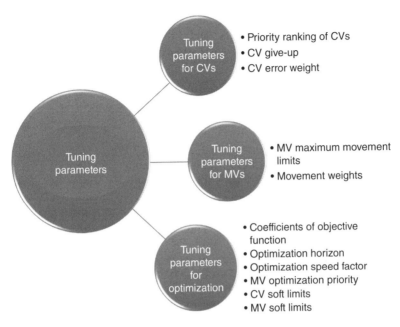

Figure 11.1 Different tuning parameters

In prediction modules, current MV, DV, CV values are given as input from plant DCS. Also, future MV values are given as input from dynamic control module. With these inputs, prediction modules predict the steady state.

In steady-state operation modules, with the help of LP and QP programs, MPC produces consistent MV and CV steady-state targets. The inputs of these modules are:

- MV, CV, high, and low limits as set by the operator
- Different tuning constants (explained later) as set by control engineer or process engineer
- Cost of manipulated variables (economics) as set by process engineer
- Steady-state prediction results from prediction module

In dynamic control modules, these MV, CV targets are input along with MV, CV limits and tuning constants. Dynamic modules then calculate future MV moves, and these moves are downloaded to DCS for implementations. In this way, MPC drives the process to the most economical zone.

11.4.1 Tuning Parameters for CVs

It is important to understand that there are no universally accepted tuning parameters. Different vendors have different tuning parameters and implement the dynamic strategy in different ways. However, the concept of major tuning parameters is the same, even though vendors have different names for it. It is important to understand the concept of tuning parameters rather than get lost in their names and intricate details.

Let us concentrate on steady-state optimization modules, where LP/QP programs are used to find feasible solutions. A feasible solution is defined as a solution where all MV and CV steady-state targets are at or within their high–low operating limits. These high–low operating limits are set by console operators. In fact, MPC always obeys MV operating limits and never violates them. However, for SV steady-state targets, sometimes it has to be violated for one or more CVs when MPC could not find a feasible solution.

Steady-state optimization works in two stages as per infeasibility check. In the first stage, it checks feasibility and finds out feasible solutions by solving a sequence of LP and QP programs. If no feasible solution exists, it temporarily relaxes the least important limits in a consistent manner. When MPC finds that all the CVs cannot be maintained within their limits at steady state, due to insufficient degrees of freedom, then it tries to sacrifice the least-important CV first and allows it to cross the limit temporarily. This is called CV give-up. The control engineer has to set the priority of CVs (i.e., set them ascending order from most important to least important CV, as per the process knowledge). MPC algorithms will give up CVs as per their priority. That is, it will try to give up least important CVs first, rather than the most important CV. This constitutes the first set of tuning parameters of CV. Different vendors has implemented the concept in different ways in their algorithm, but the concept is more or less the same. Some vendors set it as priority rank of CV and some set it as CV give-up in engineering unit.

11.4.1.1   Methods for Handling of Infeasibility
Different vendors implement the CV give-up strategy in slightly different ways, but the main concept remains the same. Priority ranking of CVs and CV give-up are two different strategies implemented by two different vendors doing almost the same job.

11.4.1.2   Priority Ranking of CVs
In this method, CVs are rank from 1 to 1,000 scale and the ranks are relative to other CVs. Limits are checked for feasibility in order of increasing rank (i.e., rank number 1 is more important than rank 10). So when optimization algorithm must sacrifice one CV, it will sacrifice CV with rank 10 and allow it to drift the limit temporarily and try to control the higher-ranked CV. Different ranks can be set for high and low limits of a particular CV.

11.4.1.3   CV Give-Up
CVs give-up values are based on the importance of keeping the CV within constraints. The smaller the give-up, the more important is the CV, the more the controller attempts to minimize the error for CV. Give-ups are relative to each other. The objective function in this stage minimizes the amount of give-up in each unique constraint. It is important to know that CV give-up has no effects when there are sufficient degrees of freedom to bring all CV errors to zero, which is true most of the time. The only purpose of give-up is to influence steady-state error trade-off when it is physically impossible to bring all errors to zero.

Second, complexity arises concerning how to decide which CV to give up and for how long if more than one CV has equal rank or priority. For a simple two-CV case, this essentially means both the CVs have equal priority. Then how much deviation from limits are to be sacrificed in each CV. Again, different vendors has implemented the strategy in different ways.

11.4.1.4   CV Error Weight
There is a low- or high-limit violation weighing factor for the CV relative to other CV. Increase the value to make the controller more sensitive to low- or high-limit violations

of this particular CV. Since different CVs have different engineering units, it is important to balance the size of constraint violations. This weight represents a constant level of concern about that size of the violation and compensates for differences in engineering units. In other words, if a CV is outside its most important limit, this error should specify how far that CV could be in violation before the operators becomes concerned. Note that this weight applies only to violation of multiple limits of equal rank or priority. When there is infeasibility, the controller maintains the best possible compromise. The best compromise is defined as minimizing the weighted sum of squared CV error, where a CV error is the amount that the CV is away from it set point or outside a limit.

Minimize

$$\sum_i weight_i^2 . error_i^2$$

where i is the CV index.

11.4.2 Tuning Parameters for MVs

There are two major MV tuning parameters, namely, MV movement limits and MV weight.

11.4.2.1 MV Maximum Movement Limits or Rate-of-Change Limits

This parameter essentially means maximum amount in engineering units that this MV can be increased or decreased by the controller in a controller execution. Some vendors refer to it as MV maximum movement limit and some vendors call it rate-of-change limit. Concept-wise, both are the same.

In basic MPC algorithm, once a steady-state target of MVs is evaluated, MPC can aggressively move the MVs and achieve those targets. However, very aggressive MV movement may cause disturbance in the process and can destabilize the whole process. Instead of that, approaching MV targets with small steps is surely a better approach. This will slowly move the process to its economic optimization zone. These limits are the maximum move that the controller can impose on an MV in a single control interval. Sometimes this is called MVs rate-of-change limit. Separate rate-of-change limits can be set for positive and negative changes. Setting very high or low values of rate-of-change limits is not effective.

Rate-of-change limits prevent the controller from making excessively large changes when an abnormal event occurs. This gives the operator a chance to intervene.

If a rate-of-change limit is set so small that the controller is hitting it much of the time, this takes away some freedom for the controller to determine the most stable and robust trajectory to use in correcting CV errors.

11.4.2.2 Movement Weights

MV movement weights are used to encourage, or discourage, controller action on particular variables. When it is desired that an MV move less than others, or not to move at all unless necessary, then a movement weight is applied. The movement weight penalizes movement of the MV, and influences the controller's choice of alternate MV moves.

This movement weight factor inhibits movement of this MV at the cost of greater total MV movement. Larger values reduce tendency to use this MV when there are sufficient degrees of freedom. This factor is relative to other MVs.

As CV engineering give-ups encourage the resolution of particular CV errors at the expense of other CV errors, MV movement weights discourage the movement of particular MVs in resolving CV error, which results in larger movement of other MVs.

All MVs are assigned an initial movement weight of 1.0, which is a neutral weighting (this number may vary for different vendors). As, for example, if control engineers set the movement weight of MV1 to twice that of MV2, the controller moves MV1 half as much as MV2, all other conditions being equal.

MV movement weights are the handles to prioritize and the different MVs to control a particular CV. One can force the controller to move particular MVs by assigning much larger weights to the MVs control engineers prefer to move very little or not at all.

If the control engineer gives MV1 a weight of 5, for example, and MV2 a weight of 1, the controller tends to leave MV1 alone, moving MV2 instead so long as MV2 is not constrained and can affect the CVs that need to be changed. If MV2 hits a constraint, only then does the controller move MV1, and only so much as to achieve the control and economic objectives.

In normal operation, usually the controller minimizes MV movement whenever possible, while still meeting both the operating and the economic objectives.

When there are more MVs than are required to meet the objectives, the controller spreads the total MV movement across the MVs to minimize the sum of the squared changes. The controller minimizes the sum of the squared changes of the MVs, with each change multiplied by the movement weight for the MV.

So, in offline tuning, appropriate MV movement weights can be set, depending on the priority of the MVs and those weights can be changed by trial and error method until a satisfactory dynamic performance is achieved.

11.4.3 Tuning Parameters for Optimizer

11.4.3.1 Economic Optimization

This section explains how MPC uses an objective function for optimizing control, and how the optimization horizon and the CV and MV soft limits influence optimization calculations.

In many applications, the control requirement to keep all variables inside constraints does not use up all the degrees of freedom available to the controller. Even when there are more CVs than MVs, there can be extra degrees of freedom if a number of the CVs have ranges rather than set points, which is often the case. Control engineers can put the extra degrees of freedom to good use by defining an objective function so that the controller optimizes some aspect of the process, in addition to controlling it.

To understand how MPC optimizes a process, some basic terms need to be understood fully:

- *Objective function:* What control engineers want the controller to accomplish after the control objectives are met (improve product spec, increase product throughput, lower utility costs).
- *Degrees of freedom:* The number of MVs not at a limit minus the number of CVs that either have set points or are at or outside limits. The controller chooses MV values so as to minimize the number of CVs that are away from set point or that are outside limits.

- *Optimization horizon:* How fast the controller must bring the objective function to an optimal value.
- *Soft limits:* Offset from high and low limits to protect future freedom for moving MVs, and for leaving room in CVs for unexpected disturbances.

11.4.3.2 General Form of Objective Function

Vendors have different forms of objective functions implemented in their control algorithm. However, to understand the concept, the following objective function is considered for illustration purposes.

The objective function is a linear or quadratic function of any or all of the CVs and MVs. The general form of the objective function is:

$$minimize\ Obj = \sum_i p_i CV_i + \sum_i q_i^2 \left(CV_i - CV_{0i} \right)^2 + \sum_j p_j MV_{ij} + \sum_j q_j^2 \left(MV_j - MV_{0j} \right)^2 \quad [11.1]$$

where

p_i are the linear coefficients on the CVs
p_j are the linear coefficients on the MVs
q_i are the quadratic coefficients on the CVs
q_j are the quadratic coefficients on the MVs
CV_{0i} are the desired resting values of the CVs
MV_{0j} are the desired resting values of the MVs

To maximize rather than minimize the objective function, multiply each term by −1 (minimizing the negative of something is the same as maximizing it). The controller minimizes the objective function (or maximizes the negative of it), subject to keeping all CVs within limits or at set point, and all MVs within limits

11.4.3.3 Weighting Coefficients

It is important to understand that weighting of the linear objective coefficients is relative. A larger absolute value relative to other CV and MV coefficients emphasizes that variable. A "−100," for example, emphasizes 10 times the desirability that "−10" attributes, so far as the optimizer is concerned. And 100 attributes 10 times the cost of 10 attributes.

11.4.3.4 Setting Linear Objective Coefficients

The linear objective coefficients are the knobs for maximizing operating profit. Here is how control engineers set the coefficients:

- Set the linear coefficients (p_i *and* p_j) of all CVs and MVs that are products to the negative of the product value. For example, if CV2 is a product stream with units of kg/hr and the product has a value of $1.2/kg, set b2 = −1.2. This assumes that the control engineer want the units of the objective function to be $/hr.
- Set the linear coefficients of all CVs and MVs that are feeds to the feed cost.
- Set the linear coefficients of all CVs and MVs that are utility streams to the utility cost.
- Set all other objective function coefficients to zero.

Setting limits rather than set points lets the optimizer determine where to drive the process.

For example, rather than putting a set point on a feed rate, control engineers can set an upper limit based on the pumping limit or on feed availability and let the optimizer determine the optimal amount of feed.

In this case, the optimization typically increases feed until one or more constraints are hit and then rides those constraints. Control engineers, therefore, have the responsibility to include in the controller all the constraints that might be hit under varying operating circumstances.

In most cases, product quality spec is one of the major constraints a controller has to obey.

With the optimizers configured to maximize operating profit, it is good practice to give limits rather than set points to product qualities. The advantage of configuring a limit rather than a set point is that the controller has more freedom to correct control errors. Remember, limits allow the controller more freedom to optimize operations and provide robust control.

Optimization, by definition, ensures that there is no unnecessary giveaway of valuable product due to a quality being better than spec. The optimization pushes qualities to their limits unless some (unusual) constraints prevent this.

11.4.3.5 Optimization Horizon and Optimization Speed Factor

There are two optimization-tuning parameters that must be set during offline simulation.

Different vendors has different tuning parameters to set the optimization horizon and speed factor. However, for sake of understanding following tuning parameters are explained.

The *optimization horizon* specifies how fast the controller must bring the objective function to its optimal value. This is set independently of the error correction horizons on the CVs.

Control engineers typically want to set the optimization horizon considerably longer than the error correction horizons. This avoids the occasional exposure of letting optimization compromise the robustness of the controller.

11.4.3.6 Optimization Speed Factor

Different vendors have different ways to set the optimization speed, which vary from vendor to vendor. Control engineers set the optimization horizon indirectly by setting the optimization speed factor. The default setting for the optimization speed factor is 1.0, which results in an optimization horizon approximately six times the CV overall response time.

The CV overall response time is defined as the average of the longest CV response time and the average CV response time. Setting the optimization speed factor to zero turns the optimizer off, which turns off the objective function. CV and MV objective coefficients, then, have no influence on the direction of the process.

11.4.3.7 MV Optimization Priority

Vendors have different ways of setting the MV optimization priority, and it varies vendor to vendor.

A prioritization scheme for accelerating the rate at which high priority manipulated variables approach their optimal steady state value. Three levels to choose from: 1 NORMAL, 2 HIGH, 3 HIGHEST. By defaults, it is 1 NORMAL.

11.4.4 Soft Limits

In addition to the above tuning parameters, setting of soft limits is another optimization tuning parameter.

In addition to the hard-set control limits, control engineers can use soft limits on the CVs and MVs. As shown in Figure 11.2 shows the basic concept of soft and hard limit.

11.4.4.1 How Soft Limits Work

Control engineers do not set soft limits directly. Rather, control engineers set the amount that the soft limit is inside the control limit. Soft limits are respected by the optimization of the economic objective function. In other words, the optimum value is subject to all variables being inside their soft limits. Soft limits are ignored for the control purposes of keeping the CVs inside their control limits or at set point.

11.4.4.2 CV Soft Limits

Control engineers can set a soft limit on a CV slightly inside the control limit to provide a buffer so that disturbances of normal magnitude do not bump the CV outside the control limit.

If the optimization limit and the control limit are instead both set equal to the actual spec on the CV, optimization then tends to keep the CV at the spec value. Consequently, even a small disturbance can push the CV out of spec until the controller reacts.

If control engineer configure the soft limit some small amount inside of the actual spec, optimization cannot push the CV all the way to the spec. Optimization, in fact, tends to bring the CV back to the soft limit if the CV is outside the soft limit but inside the control limit. This action proceeds at the speed determined by the optimization horizon and provides a buffer. If a large disturbance pushes the CV outside the control limit, the much more aggressive error correction horizon comes into play, and the CV is quickly brought back within the control limit.

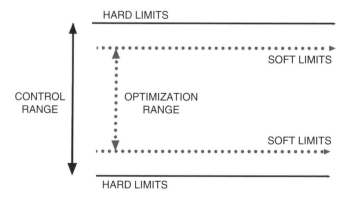

Figure 11.2 Hard and soft limits

11.4.4.3   MV Soft Limits

Control engineers can set a soft limit on an MV slightly inside the control limit to ensure that the MV is available to correct disturbances in either direction most of the time. If the controller moves the MV outside a soft limit to correct CV errors (it cannot move the MV outside its control limit), then the optimizer adjusts other MVs to gradually bring this MV back within its soft limit.

Keeping MVs slightly inside their control limits can improve controller robustness by giving the controller more freedom to correct CV errors.

11.5   Different Steps to Build and Activate Simulator in an Offline PC

The actual steps to build and activate a simulator in an offline PC vary from vendor to vendor and can be found in vendor manuals. To understand the concept, some basic steps are given below, without giving actual details. Actual details are available with every MPC software package and are easy to implement with some clicks:

- Step 1: Select initial tuning values of MVs and CVs. The initial values will be used in simulators and should be given judiciously. This essentially means selecting some good guess values of move suppression, maximum MV size, and CV priority to start with.
- Step 2: With these initial tuning values, create online tuning files. This file is a database that contains information required to define a MPC controller to the MPC software. It usually contains the model of the process, LP matrix, various constants, and tolerances.
- Step 3: Create a simulation initial conditions file. Information requirements vary from vendor to vendor, but some basic tuning parameters are shown in Table 11.1 and Table 11.2.
- Step 4: Build the offline controller: The process simulator is used to provide a process that may be controlled by the controller we are going to build. Select the BUILD option on the menu bar and select the "Process Simulator…" submenu item (menu

Table 11.1 Simulation Initial Condition File for MVs

For MVs	Remarks
Current value	Give realistic value, refer current value from DCS.
Status	Keep it default.
Upper and lower limit	Use realistic value after discussing with plant operator.
Maximum movement limit	Use initial realistic value based on the guidelines. These values need to be modified during simulation.
Movement weight	
LP weights coefficient	
MV optimization priority or weights	
MV soft limits	

Table 11.2 Simulation Initial Condition File for CVs

For CVs	Remarks
Current value	Give realistic value, refer current value from DCS.
Status	Keep it default.
Upper and lower limit	Use realistic value after discussing with plant operator.
Priority ranking of CV	Use initial realistic value based on the guidelines. These values need to modify during simulation.
CV give-up	
CV error weight	
LP weights coefficient	
CV soft limits	

item may vary for different vendor). Enter the execution frequency. Note that it MUST match that of the controller. View the simulated CV view normally available in offline MPC software.

- Step 5: Design different test to carry out in the simulator: Think of different tests, which need to be performed in offline simulations. Tests should be designed such that different dynamic features of online MPC controller can be evaluated and different tuning parameters can be adjusted to get an optimal acceptable performance. The number and types of test varies plant to plant, MPC software to software, but some basic examples (not limited to) are given in next section.
- Step 6: Carry out those tests in simulator and evaluate how controller perform in a dynamic environment. Keep on changing tuning parameters until a good acceptable dynamic performance is achieved.

11.6 Example of Tests Carried out in Simulator

Let us take a simple example of a binary distillation tower to understand the concept of offline simulation test and how to perform them. The distillation tower is drying column where moisture is removed from glycol water mixture. Water being a low-volatile compound, essentially vaporize and remove from the top. Bottom of drying column is heavy glycol with little moisture (700 ppm max).

The distillation tower with the controlled scheme is shown in Figure 2.4.

As per the Figure 2.4 and Table 11.3 and Table 11.4, there are six control variables and three manipulated variables. Feed flow is considered as disturbance variable. Low and high limits of CVs, MVs, and DVs are given in Table 11.3 and Table 11.4.

11.6.1 Control and Optimization Objectives

Let us assume there are four control and optimization objectives:

1) Maintain all the CVs within their limit.
2) Maximize moisture content at bottom (CV1) but always obey the high limit (700 ppm moisture max).

Table 11.3 Controlled Variables with Their Limits for Simulation Studies

Controlled Variables	Description	Low and High Limit	Current Value (for Simulation Study)
CV1	Bottom moisture content is measured by a continuous online analyzer, ppm	200–700	600
CV2	Top stream purity i.e., glycol content, ppm	1000–2000	1800
CV3	Top temperature, °C	55–60	59
CV4	Bottom level, %	20–90	60
CV5	Flooding %, calculated by soft sensor, %	20–80	70
CV6	Top level control valve opening, %	20–90	50

Table 11.4 Controlled Variables with Their Limits for Simulation Studies

Manipulated & Disturbance Variables	Description	Low and High Limit	Current Value (for Simulation Study)
MV1	Reflux flow rate, MT/hr	25–35	30
MV2	Reboiler steam flow rate, MT/hr	45–55	50
MV3	Column bottom flow rate, MT/hr	70–80	78
DV1	Feed flow rate, MT/hr	110–120	115

3) Maintain top temperature below the maximum limit (60 °C max).
4) Minimize steam consumption in reboiler (MV2).

CV priority: Priority of CVs in decreasing order of importance are as follows: CV1 (most important), CV2, CV6, CV3, CV5, CV4 (least important). This priority ranking is for illustration purposes only, priorities would be assigned for each plant on a case-to-case basis.

With this setup, we need to design different tests to study how the controller perform in a dynamic environment. Select initial value of different tuning parameters such as CV give-up, CV error weight, maximum movement limit, and movement weight as per above priority following the guidelines just given.

The following tests illustrate how this works in various scenarios.

11.6.1.1 Test 1

Objective: See how the controller performs to control different CVs when limit was changed in one of the main CVs.

Test setup: Reduce the limit of one main CV and see how controller performs. Controller performance can be seen from the trends of CVs and MVs available in the simulator. For example, reduce the higher limit of CV1 (moisture content at bottom) from 700 ppm to 400 ppm. Essentially, this means tighter water specs at drying column bottom. Physically, this means steam has to increase in reboilers to vaporize more water. As the process is highly interactive, increasing steam will increase top temperatures also. MPC being a multivariable controller, it will adjust reflux and steam so that it will maintain the top temperature within its high limit.

Observations: After reducing the upper limit of CV1, run the simulation and see the steady-state values of all CVs and MVs. To assess the dynamic behavior of the controller, see the different trends of MVs and CVs in the simulator window.

See the following critically:

- Is steam flow increased too rapidly?
- Can controller reduce the moisture content to 400 ppm while maintaining the top temperature within its high limit?
- Is reflux flow increased too aggressively?
- Can controller maintain all the CVs within their limit?

Recommendations: Follow the guidelines explained later if controller performance is not satisfactory.

11.6.1.2 Test 2

Objective: Check how the controller handles insufficient degrees of freedom.

Basically, the number of degrees of freedom is the number of MVs not at a limit minus the number of CVs that either have set points or are at or outside a limit. The controller chooses MV values so as to minimize the number of CVs that are away from set point or outside limits.

As long as the degrees of freedom are zero or positive, all CV constraints can be satisfied. If the degrees of freedom become negative, it is physically impossible to keep all CVs at set point or within a range.

Test setup: Controller can control as many CVs at a constraint as there are MVs available. So if we have six CVs and three MVs we can control all six CVS if they are within their range but only three if they are hitting constraints or at set point.

Now we want to make all six CVs as set points. We make the high and low limit of all the CVs at its current value. This will make all six CVs as set points. Notice that we don't achieve the set points because we only have three degrees of freedom!

Observations: Now run the simulations and see the steady-state value and CV, MV trends. Controllers should able to maintain three important CVs as per their priority—namely, CV1, CV2, and CV6. It should give up other CVs, CV3, CV4, and CV5. Observe whether the controller is performing as intended.

11.6.1.3 Test 3

Objective: Check how the controller handles the CVs when some CV is made more important than another.

Test setup: Now CV give-up is set such that the intended priorities are made as described below.

Priority-wise, CV1 > CV2 > CV6 > CV3 > CV4 > CV5

We make the high and low limit of all the CVs very narrow and just ±1% around its current value.

Observations: Now run the simulations and see the steady-state value and CV, MV trends. Controller should able to maintain three important CVs as per their priority namely CV1, CV2, and CV6. It should give up other CVs, CV3, CV4, and CV5, if at all it has to give. See the controller is performing as intended.

11.6.1.4 Test 4

Objective: Sometimes we don't want to be too aggressive on a particular MV. MV weights allows us to do it. The objective of this test is to observe how controllers move different MVs when a set point change in CV is done. Another objective is to set different values of MV weights, so that the MV movements are as per the operation engineer.

Test setup: Make a big change of set point in one of the major CVs and observe how the controller moves different MVs to reach that set point. Do it for all the CVs.

Observations: Now run the simulations and see CV and MV trends. Observe if MV movements are as per their priority. Increase MV weight for particular MV if operation engineer feel that the particular MV movement should be smaller or restricted. Rerun the simulation after increasing the MV weight. That MV is still used, but smaller moves would results. See that the controller is performing as intended. Run the simulations many times to cover all MVs. The final results should be to set different values of MV weights, so that the MV movements are as per the operation engineer,

11.6.1.5 Test 5

Objective: The objective of this test is to observe how controllers move different MVs and control CVs when a big change in DV is done. Another objective is to set the speed for rejecting feedforward disturbances, so that the MV movements are as per the operation engineer.

Test setup: Make a big change in major disturbances in process i.e. in DV (say cooling water temperature or feed flow to the process) and observe how controller moves different MVs to maintain all CVs in their desired limits. Do it for all the DVs.

Observations: Now run the simulations and see CV, MV trends. Controller should be able to maintain all the CVs in their desired range in spite of big disturbance entered into the system. Evaluate the performance controller from that point of view. In some vendor software, control engineer can change the speed of feedforward disturbance rejection. Change that speed if controller performance is no satisfactory. Rerun the simulation. See the controller is performing as intended. Run the simulations many times to cover all DVs.

11.6.1.6 Test 6

Objective: The objective of this test is to observe how controllers and optimizers move the process to a most economic zone when degrees of freedom is available. Another objective to set the different LP coefficient in LP/QP objective function so that moving toward the economic zone is as per design intent.

Test setup: We will see how with optimization we can move CVs or MVs in a desired direction to optimize the plant. A negative value on coefficients of CV1 will maximize moisture content at the bottom.

Observations: Now run the simulations and see CV, MV trends. Controller should be able to maximize CV1 (bottom moisture content) while maintaining all the CVs within their prescribed limits. By doing so, it will able to minimize steam in reboiler (MV2). Evaluate the performance controller from that point of view. Change the negative value of the LP coefficient of CV1 if controller performance is not satisfactory.

Rerun the simulation. See that the controller is performing as intended. Run the simulations many times to cover all optimizing CVs and MVs. Sometimes by setting different LP coefficients of two competing MVs, their optimization movement can be prioritized as per their cost. Configure different test for that so that appropriate LP coefficients can be set. The final results should be to set different values of LP coefficients of CV and MVs, so that the controller optimization priority movements are as per the operation engineer.

11.6.1.7 Others Tests

Many other tests can be configured as per the demand of a particular process and controller objectives. This varies from plant to plant. How to implement this tests and which parameter to tune also vary from vendor to vendor. Other tests are also carried out. Tests 7 and 8 investigate the effect of optimizer speed. Test 9 observes the effect of MV rate-of-change limit on CV error. Test 10 observes the effect of MV rate-of-change limit, MV movements on CV error, and CV give-up.

It is important to understand the concept that nothing is free and everything is a trade-off. Tuning tight for a particular CV means good control for that CV but inferior control for other CVs. The overall objective of offline simulation is to tune different parameters in controllers and optimizers so that an acceptable offline performance and reasonable trade-off can be achieved.

11.7   Guidelines for Choosing Tuning Parameters

Initially, default values (as prescribed by different vendors in their software package) or some smart initial values (judicially selected as per guidelines) for these tuning parameters are chosen to start the simulation. Offline simulation will be used to evaluate these values and will be modified to bring about the desired control performance. The main purpose of the offline simulation is to adjust these values so that an acceptable optimal dynamic performance of the controller can be achieved.

Tuning strategy for MPC is discussed in detail in Al-Ghazzawi (2001); and Ali (2003). Tuning guidelines for DMC controller are given in Dougherty (2003). A review of MPC tuning methods is discussed in Garriga (2010).

11.7.1   Guidelines for Choosing Initial Values

Initial CV give-up values should be chosen in such a way that if the controller has a choice of giving up on a key composition quality or operational safety or flooding, it should always control and give priority to safety first, constraint limit next, and quality limit last. So, CV give-up for safety-related variables should be made smallest (meaning, of highest importance), variables representing equipment or process constraints should be higher, and finally quality-related variables should be the highest. Note that these are general guidelines to set initial values to start with. These values can be changed based on judgment on a case-by-case basis.

Initial MV movement weight values: For MVs that are to be freely moved, set the MV movement weight lowest (say, 1). An example of this might be a furnace fuel-flow controller set point, where the ability to rapidly decrease the set point in response to a violation of tube metal temperature high limit is essential.

Increase MV movement weight for MVs that should be moved very slowly (say, 10). An example would be a distillation column pressure set point.

The remaining other MVs should be given MV movement weight in between (say, 5). Reflux flow and reboiler steam flow may fall in this category. It is noteworthy to mention here that the above guidelines are used to choose initial values. The effectiveness and applicability of these values should be judged during offline simulation. These values can be modified if the performance is not satisfactory.

11.7.2 How to Select Maximum Move Size and MV Movement Weights During Simulation Study

Maximum move size should be set large, and it should not be used to restrict the MV movement. MV movement weights should be used for that.

Normally, it is good practice to run a steady-state offline simulation with all MV and CV limits in their realistically normal value. During simulation testing, introduce a large feed-flow change (assuming feed flow is a DV for the simulation test case) or give large input in any DV of the process. DV magnitude should be approximately equal to the maximum expected on the real process. This will force the controller to take large MV steps to nullify the effect of the disturbance. Check how aggressively the controller is moving the MVs. MV movement weights are then adjusted to keep MV variable moves below the maximum allowable value specified by the operating personnel.

Detailed guidelines on how to choose tuning parameters are given in Chapter 13. Those guidelines are valid for offline simulations also.

With this knowledge, tuning parameters are modified by trial and error in offline simulations until an acceptable optimal dynamic performance is achieved.

It is extremely necessary to understand the limitations of offline simulation where the controller is being tuned under "perfect" simulated process conditions. The effects of unmeasured disturbances and measurement noises are not being taken into account at this stage of the project. The goal of the offline tuning exercise is to simply obtain some initial values of the tuning parameters so that the controller can be deployed. These values will be further adjusted during online tuning after controller commissioning.

References

Al-Ghazzawi, A., Ali, E., Nouh, A., & Zafiriou, E. (2001). On-line tuning strategy for model predictive controllers. *Journal of Process Control*, 11(3), 265–284.

Ali, E., & Al-Ghazzawi, A. (2003). On-line tuning of model predictive controllers using fuzzy logic. *Canadian Journal of Chemical Engineering*, 81(5), 1041–1051.

Dougherty, D., & Cooper, D. J. (2003). Tuning guidelines of a dynamic matrix controller for integrating (non-self-regulating) processes. *Industrial & Engineering Chemistry Research*, 42(8), 1739–1752.

Garriga, J. L., & Soroush, M. (2010). Model predictive control tuning methods: A review. *Industrial & Engineering Chemistry Research*, 49(8), 3505–3515.

Tuning strategy for MPC is discussed in detail in Al-Ghazzawi 2001; Ali 2003. Tuning guidelines for DMC controller is given in Dougherty 2003. A review of MPC tuning methods is discussed in Garriga 2010.

12

Online Deployment of MPC Application in Real Plants

12.1   What Is Online Deployment (Controller Commissioning)?

Commissioning of the controller means connecting the MPC controller online with the plant DCS and allowing it to take control of the plant. This is the final and most important step. Commissioning of the controller at the online platform in a running plant is a critical job and need 24-hour coverage. Care should be taken so that any mistake during commissioning does not lead to plant shutdown or plant upset. The most common mistake is to read and write to a wrong tag in DCS, inducing process fluctuation due to MPC poor model or tuning (see Wang 2002; Bowen 2008; Clarke 1988; McDonald; 1987).

12.2   Steps for Controller Commissioning

Different vendors have different methodologies to commission the controller. However, the basic steps are the same:

1) Set up the controller configuration and final review of the model.
2) Build the controller.
3) Load operator station on PC near the panel operator.
4) Take MPC controller in line with prediction mode.
5) Put the MPC controller in close loop with one CV at a time.
6) Observe MPC controller performance.
7) Put optimizer in line and observation of optimizer performance.
8) Evaluate overall controller performance.
9) Perform online tuning and troubleshooting.
10) Train operators and engineers on online platform.
11) Document MPC features.
12) Maintain the MPC controller.

12.2.1   Set up the Controller Configuration and Final Review of the Model

This is a preparation step of controller commissioning. The following actions are normally performed.

Multivariable Predictive Control: Applications in Industry, First Edition. Sandip Kumar Lahiri.
© 2017 John Wiley & Sons Ltd. Published 2017 by John Wiley & Sons Ltd.

It is a prerequisite to ensure that all the issues / findings of offline simulations are properly addressed and tuned accordingly. If offline simulation does not behave like a real process, a complete revalidation of the model is necessary. Revisit raw data, modify the model curve, and change CV, MV, and DV—the relationships should be tried and later on run in a simulator. Once satisfied, freeze the model curves. The main aim is to generate a robust, reliable model that is able to control the process.

Now, solidify the controller specifications, including adding proper high–low limits and ramp limits. It is important to get buy-in from operators.

Check the PID loops in DCS. Ensure that they are at their desired position. Break the cascade mode where it is needed as per final functional design.

12.2.2 Build the Controller

The next step is to *configure the online controller.* The details of this task are specific to the type of process control computer, process control database, and distributed control system (DCS), which are connected to the controlled process.

This step involves building the appropriate database tags, building operator and engineer displays, and loading the controller configuration file (CCF) to connect the database tags to the actual controller.

The program should be tested for proper data transfer to and from the process control database. Proper execution of the controller should be verified as well.

Building the controller means connecting and mapping MPC variable to process control system (PCS) / Distributed control system (DCS) parameters.

Define the variable transformation and calculation. Define external communications like external target, communication services, and so on.

Normally, the term *build* is used to define the communication relationships between the controller computer program and DCS. Build basically convert MPC model file to controller configuration file (*.ccf). Normally model file contains process information in terms of process model, which is used to define process relationship between independent and dependent variable. The controller configuration file (*.ccf) created in build process contains the link information or communications relationships between the controller and DCS. Special handling and logic is also defined in CCF. It contains the address of read/write destinations. Sometimes a separate module called *MPC controller interface module* is used to represent online memory of CCF. Controller contexts are used to map DCS locations for values of parameter with external connections. All these tasks and terminology varies from vendor to vendor. Figure 12.1 explains the complete interface between MPC computer and DCS. This is just for illustration purposes, actual interface may vary vendor to vendor, DCS to DCS.

At the end of this task, the model and configuration file is loaded in MPC online computer.

12.2.3 Load Operator Station on PC Near the Panel Operator

This facility is already available in most of the MPC software. Normally, a separate computer is kept as an operator display station where operators can see the various parameters of MPC online in real time. Install the operator display program in the operator station and explain the various terms to the operators so that they can follow and change the limits as required in real-time operation. Normally, installation

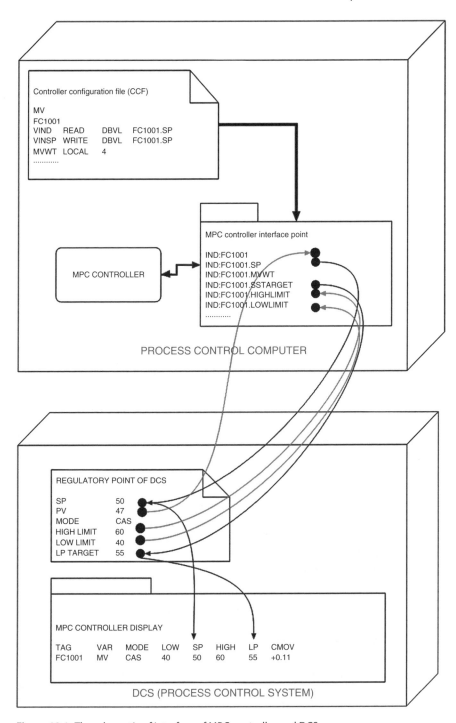

Figure 12.1 The schematic of interface of MPC controller and DCS

guidelines vary from vendor to vendor and step-by-step installation procedures can be found in a vendor manual. Readers are advised to refer to the user guide / manual of the respective MPC vendor for details.

Check all other software-related issues (if any) and ensure that software are working properly.

12.2.4 Take MPC Controller in Line with Prediction Mode

At this point, online controller commissioning is performed. Control engineer will use vendor software online module to start and maintain the controller, and monitor the controller graphically to make any necessary changes. It is advisable to read the vendor manual for details.

The MPC online control program is first run in *prediction mode* for a day or two to verify proper program functioning. Check the accuracy of the model. In this mode, the controller performs all control calculations, but the moves are not sent to the process.

Controller in prediction mode can only read the PV value from DCS and predict the steady state value of CV. In prediction mode, it cannot write back the set point to PID controller in DCS. The idea of taking the MPC controller in line with prediction mode is to see that the controller is reading the correct value from the DCS tag. It ensures that the all tags are mapped correctly between MPC and DCS. If there is any error in that, it will provide a chance to correct it without affecting the process, as controller is only reading the tag (one way) and not writing back anything. This is the main objective to putting the controller in prediction mode. During this time, informal operator training takes place. A brief operator guide is provided, specifying the objective of the MPC controller, the variables that it considers, the operator displays to be used, and the procedure for turning the controller on and off.

The control engineer works with the operators to explain how to interpret the information on the operator displays and how to use the displays to interact with the MPC controller.

It is normal practice to first install and commission the controller and then commission the optimizer. In some MPCs, commercial software controller and optimizer can be install separately. Readers are advised to refer to the manual of the respective MPC vendor for more details.

General guidelines are as follows:

- Switch off the optimizer by setting the optimizer speed to zero.
- Commission the controller as per guidelines written in the vendor manual.
- Run the live prediction with controller in OFF mode.
- Ensure that the controller is reading the correct CV from DCS. Compare the values of CV in MPC controller and DCS. They should be the same.
- Observe the unforced prediction of CV by looking at its trend. See the historic trend of unbiased prediction and compare it with CV trends. Two trends should move in the same directions. This is an indirect verification of goodness of the dynamic model.
- Check the MPC shedding mode. This is the time before the controller sheds to its shed mode. Set time out to two times control runtime frequency.

12.2.5   Put the MPC Controller in Close Loop with One CV at a Time

Start the controller in warm mode and monitor the CV and MV trends. Once prediction mode steps are accomplished, the actual commissioning takes place. All manipulated variable limits are pinched in very near the current set point values so that only very small control moves are allowed.

The MPC controller is then turned on. Once the controller runs, calculates moves, and implements these moves on the process, the control engineer must make sure that the calculated moves are actually implemented at the regulatory control level.

Initially attempt to put one CV along with its one MV and see its behavior as an SISO controller (Figure 12.2).

Ensure that the controller is reading and writing the correct CV and MV in DCS.

Keep an eye on the overall process so that any upset during commissioning due to some mistake can be dealt with quickly.

Extra precautions should be taken for those CV and MV, which are initiators of shutdown systems to avoid unnecessary plant shutdown. Then put another MV and so on, and allow MPC to use multiple MVs to control the CVs (one at a time, as shown in Figure 12.2).

Initially narrow down the MV high-low limit during commissioning so that the controller cannot upset the process in case of poor model or poor prediction.

12.2.6   Observe MPC Controller Performance

- Observe the control performance of MIMO controller and watch the following (one CV at a time).
- Check whether MPC is able to control the CV within its allowable range. If not, then investigate the reasons.
- Check that controller is taking corrective and preventive actions as desired or as thought of during offline simulations.
- Check whether it is utilizing the MVs as per their priority or as set by operation people.
- MV movements and step sizes are normal and not erratic, so that they can introduce fluctuations in the process. If necessary, adjust the MV weights to change the distribution of movement between the MVs.
- Check that if the controller has constraints by infeasibility (i.e., when the controller has to give up one or more CV), it should always prioritize it as follows: Safety first, constraints limit next, quality limits/softer constraints last.
- Check that the CV trajectory with set point change give good control. Tune if necessary.
- Tune the feedforward/feedback performance ratio to give good feed forward disturbance rejection (This feature is available in RMPCT but may vary for other vendor technology.)
- Keep on adding CV until all the CV and MV of the model are under MPC control.
- Fine-tune different parameters to have a smooth control

At this point, the manipulated variable limits are gradually relaxed, controller performance is evaluated, and retuning is done as required. During the controller commissioning phase, control engineers again work around the clock in the control room with the operators.

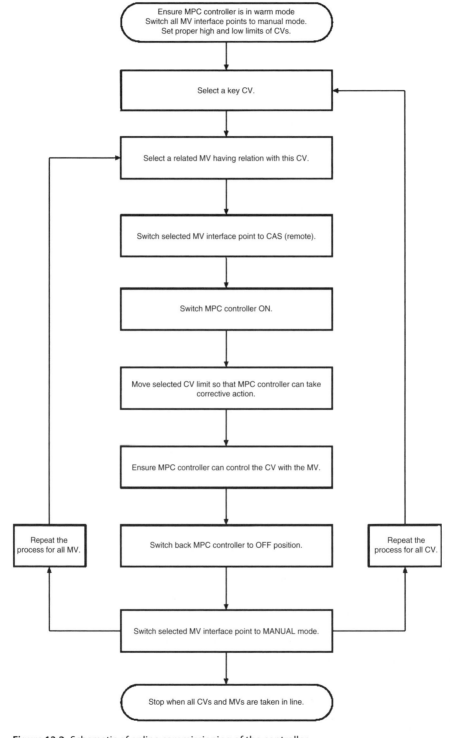

Figure 12.2 Schematic of online commissioning of the controller

12.2.7   Put Optimizer in Line and Observe Optimizer Performance

With the MPC controller ON, switch on the optimizer with slow speed. Normally this feature is available in most of the commercial MPC Controller.

Observe the optimizer targets for MVs and CVs. Evaluate whether these targets make sense from a process engineering point of view:

- Initially restrict CV and MV soft targets.
- Verify how optimizers manage infeasibility (if any) and the CVs give up as per planning.
- The job of the optimizer is to move the process into more optimum zones when degrees of freedom is available. Check whether optimizer is doing this job.
- After some time, try to increase optimizer speed. Increase optimizer speed or reduce soft limits for the good-quality models (i.e., where the CV-MV relationship is found good). However, reduce optimizer speed and increase soft limits for poorer models.
- If CV is very noisy, slower optimizer speed gives better results.

12.2.8   Evaluate Overall Controller Performance

Evaluate overall performance of controller and optimizer from the following viewpoint. Note that first priority of MPC controller is to control the process and then optimize it if degrees of freedom is available.

Evaluate whether the MPC is meeting its objective as per the following order of importance:

- Prevent violation of input and output constraints.
- Drive the CVs to their steady-state optimal values (dynamic output optimization).
- Drive the MVs to their steady-state optimal values using remaining degrees of freedom (dynamic input optimization).
- Prevent excessive movement of MVs.
- When signals and actuators fail, control as much of the plant as possible.

Evaluate the overall controller performance based on its performance on the following criteria:

- Whether multivariable controller able to maintain and control the process within reasonable time and accuracy.
- Whether correct MVs are moved to control relevant CVs. Whether MVs move as per their priority.
- Whether MV movements are in controlled way or it gets bumpy and aggressive. Simulation is used to tune move suppression.
- Whether optimizer perform its duty and able to drive the process in more economical operating zone.
- Whether optimizer able to push the process at its constraints or limit.
- How insufficient degrees of freedom is handled (i.e., how controllers give up some of the CVs, as per their priority and able to control the process in a feasible region).

After running the MPC application for three to four days and troubleshooting all the problems, work with the operator to relax MV limits. A broad MV limit facilitates MPC application to exploit the opportunity and extract benefit. A well-designed application should not have restrictive MV limits.

12.2.9 Perform Online Tuning and Troubleshooting

In this step, detailed troubleshooting is performed if the controller does not behave as it was planned or desired. Various controller-tuning parameters are adjusted accordingly:

- Observe the CV and MV trends before and after deployment of MPC.
- Evaluate whether the MPC itself induced some cyclic or noisy behavior in the process of certain CVs.
- Check whether MPC is able to reduce standard deviations of key process parameters.
- Evaluate how MPC taken care of CVs when unmeasured disturbances are entered into the process.
- Depending on the performance of MPC, various tuning parameters are changed online.

These steps are discussed in detail in next chapter.

12.2.10 Train Operators and Engineers on Online Platform

Operators and plant engineers are the ultimate end users of MPC application. It is absolutely necessary to get their support and confidence at every step of MPC implementation. If they are unhappy about the controller it will be switched off. One of the key factors for success of MPC is a well-trained operator. The following training is suggested to ensure that MPC application always receive support and confidence from operator.

It is important to arrange two to three days of classroom training, explaining all the features of MPC.

This should be accompanied with one to two days of on-panel job training, where all the online modules are explained. Each term in the operator station, how to start and stop the application, what to do in case a particular MV is bad, and so on should be explained in detail.

Sometimes MPC took actions that appear to be in contradiction to the operators' viewpoint. At that time, it is necessary to explain to the operator why MPC is taking actions in such away. Convey the idea of concerted MV moves and show them why and how a particular action is taken by the controller.

A separate three-to-four day engineer/supervisor training is needed for plant engineers who will be responsible for day-to-day MPC activities.

Delaying formal operator training until after commissioning allows the training to be done in the context of the final controller. However, it is always beneficial to get the operators and engineers involved early and at each stage of implementation process. This will allow them to follow easily and help them to keep track of why a particular action is done.

12.2.11 Document MPC Features

Documentation plays an important role due to obvious reasons. High turnover of operators, process engineers, control engineers, and MPC engineers are very common nowadays. A good documentation of all the features of MPC can be very helpful to retaining the knowledge base. It will be act as a source of important information of future generations of operators and engineers. At the end of the implementation steps, the following documentations must be created for future record.

- Functional design in detail: It should specify MV, CV, DV list, controller objectives, optimizer objectives and so on.
- Functional design intent should be specified very clearly. Why and how MVs will control CVs as per functional design should be elaborately documented.
- Model curves with gain and delay time values should be documented.
- Relevant DCS screen should be documented for clarity.
- MPC tuning screen should be documented.
- Values of all tuning parameters of MPC should be documented with proper explanations.
- Soft sensors model equations should be documented with proper explanations of all terms.
- Custom programs (if any) should be documented with explanations.

12.2.12 Maintain the MPC Controller

For any MPC application, some amount of *maintenance* is required. The most frequently performed maintenance task is monitoring the operating limits on the manipulated and controlled variables in the problem, to be sure that they are spread out to reflect the largest operating region allowed. Left unattended, these limits often get pinched closer and closer, thus limiting the controller's ability to optimize the process. The other major task involved in maintaining the controller is to keep the LP costs updated when new product values, or feed/utility costs, become available. If the calculations are captured using a spreadsheet, it is a simple matter to run the calculation when the new prices become available and enter these values into the process control database.

References

Wang, C. M., & Lei, R. X. (2002). Application of multivariable predictive control technology in atmosphere and vacuum distillation unit. *Petrochemical technology and Application*, 20(5), 321–323.

Bowen, G. X. D. F. X. (2008). Application of Advanced Process Control, RMPCT of Honeywell, in DCU. Process Automation Instrumentation, 4, 013.

Clarke, D. W. (1988). Application of generalized predictive control to industrial processes. *IEEE Control systems magazine*, 8(2), 49–55.

McDonald, K. A., & McAvoy, T. J. (1987). Application of dynamic matrix control to moderate-and high-purity distillation towers. *Industrial & Engineering Chemistry Research*, 26(5): 1011–1018.

13

Online Controller Tuning

13.1   What Is Online MPC Controller Tuning?

When an MPC application is put online, it may not perform well as its performing in offline simulations mode. Reasons are many. This may be due to unmeasured disturbances' impact the CV, too much of which model cannot capture. Another probable reasons might be that for the step test model, prediction is poor due to error in the model. Performance of MPC may not be par, due to either process behavior or controller behavior. Purpose of online tuning is to tweak various tuning parameters of MPC controllers, so that controller performance can be enhanced.

13.2   Basics of Online Tuning

When an MPC controller put online and its performance is not good as expected, the probable reasons are many. Only solution to this problem is to understand the problem and investigate the reasons. There is no shortcut on this. This is the area where the expertise and experience of true control engineer or MPC vendor comes into play. MPC controller tuning is much more complicated than simple PID controller tuning, as much more tuning parameters are involved in multivariable case. Also, MPC controller tuning is an art of delicately balancing different competing features of MPC, and it varies from vendor to vendor. Many researchers (Al-Ghazzawi 2001; Ali, E 2003; Garriga 2010; Shridhar 1997, June; Shridhar 1997) has investigated various aspects of MPC controller tuning and recommended some methodology to logically tuned the multi-variable controller. Tuning strategy for MPC is discussed in detail in Al-Ghazzawi (2001) and Ali (2003). Tuning guidelines for DMC controller is given in Dougherty (2003). A review of MPC tuning methods is discussed in Garriga (2010).

13.2.1   Key Checkout Regarding Controller Performance

Under MIMO control and handling of key constraints—for example, if controllers have a choice of giving up on a key parameters, it should always go for safety first, constraint limit next, and quality limit/softer constraints last. Please refer to the six tests performed in offline simulations. It is important to know how the controller performs in online actual environment if one or more of those test conditions appear in real plant. Ensure that controller performs as intended; otherwise, tune the parameters again with the following guidelines in next section.

Under normal operation, the controller should be able to control all the CVs within their limit, and if extra degree of freedom is available, controller should actually be controlling on the economically attractive limits, and not just sitting on the flooding or equipment constraint limits all the time. Only then will it optimize the prices and fetch economic benefit. Otherwise, it will only control the process.

13.2.2 Steps to Troubleshoot the Problem

If performance of MPC controller is not at par, investigate the root cause and troubleshoot the problem rather than jumping in to tune the controller. It is important to understand the difference between tuning problems and nontuning problems:

1) Identify which CVs the controller could not control properly.
2) Take one CV at a time. Using the model matrix, identify the MVs and DVs that are controlling this particular CV.
3) Observe historical trends of these MVs, CV, and DVs and analyze the trend very deeply. Actual CVs and unbiased CV predictions trend should be plotted together in a same plot. These two trends should move together when the CV limits are changed. It is understood that these two trends should be a little bit apart due to unmeasured disturbances that the MPC should reject in every execution.
4) Evaluate whether these two trends are far apart. If they are far apart, investigate whether this is due to some MV or DV that is missing in the model matrix that is affecting this CV but not captured by model. Maybe the particular MV impact was not thought of during functional design stage. If this is the case, add some probable MV or DV model curve in the model matrix and evaluate again the performance.
5) Evaluate if the poor CV prediction is due to model inaccuracies and model mismatch with reality. Adjust the gain and dead time biases and see whether prediction improves.

Poor MPC performance may be due to some or all of the following reasons:

- *Process behavior may be change during step test and when controller put online.* Catalyst selectivity may change, heat exchanger may foul, separation efficiency of distillation column may change due to tray or packing fouling, or the plant may be operating completely different throughput, which is far away from the throughput during step testing. Crude oil quality may change; polymer grade may change. MPC can address some of these parameter changes through its feedback route. But if the process behavior changes drastically and that is not captured by the MPC model, then it is futile to rectify it with online tuning. So, it is important to understand the above concept before jumping to change tuning parameters. Not everything can be resolved by online tuning.
- *Poor performance due to controller behavior can only be rectified by online tuning.* It is extremely important to investigate the reasons why the controller induces some disturbance or cyclic behavior in the process. Track down what is caused by the controller and what is not.
- *If a particular CV could not be controlled, then find out the reasons.* Possible reasons include (but not limited to) the following:

- Consider what happened physically. Understand what has happened both up and downstream sections and how those thing impacting the current CV.
- Determine whether the MV limits may be too much conservative. Relax those limits.
- Heavy move suppression or a low rate of change limit on some MVs ultimately makes the corrective action slower.
- Check whether priority of high-gain MVs may be lower in the tuning panel. Hence, controller could not use those MVs effectively. Maybe the controller tried to use low-gain MVs to control CV, and this could not control it in a reasonable time frame. If this is the case, then increase the priority of high-gain MVs and evaluate the performance.
- CV priority may be too low. Hence, the controller gives up this particular CV.
- The operator might not give proper CV limits, such as limits given to bottom temperature, tray temperature, and top temperature of distillation column. There is some relation between these temperature—they are not fully independent. Operators should understand that, and different temperature limits should be properly chosen so that they are relevant.
- CVs may be heavily affected by some DV that is not accounted in the model. Include that DV curve so that the controller knows its impacts.
- Analyze the future trend of CV. Find out how much future error the controller plans to resolve. Track what happens afterward. Find the discrepancy and the causes.

It is advisable not to overtune the controller. Give $3\,T$ time to controller and allow it to function. Give the controller enough time (one to two days) to settle down and take corrective and preventive actions.

13.3 Guidelines to Choose Different Tuning Parameters

Initially, default values (as prescribed by different vendors in their software package) or some smart initial values (judicially selected as per guidelines) for these tuning parameters are chosen to start the simulation. Usually, offline simulation will be used to evaluate these values and will be modified to bring about the desired control performance. In offline simulations, impacts of unmeasured disturbance variables on CVs are not usually considered. Also in offline simulations, process response is treated as simulated response. Actual process response may differ when the controller is put online. The main purpose of the online simulation is to adjust these values so that an acceptable optimal dynamic performance of the controller can be achieved.

The dynamic controller performance is judged based on the amount of error in the dependent variables from its target compared to the amount of movement in the MV. An ideal good controller will try first to minimize the CV error, maintaining all the CVs within their limit with minimum MV movement. If degree of freedom is available, it will slowly drive the process to the most economic zone. Basic algorithm of MPC aggressively minimizes error between CV targets and predictions. It is always better for some MVs to have smoother, less aggressive control actions. So there must be a balance between two conflicting objectives.

The first objective is to minimize CV error from target; the second objectives is to minimize MV movement. CVs give up and MV movement weights are used to trade off between these two conflicting objectives. It is necessary to understand the underlying concept of this trade-off before attempting to offline or online tuning.

Increasing MV movement weight will reduce the MV movement, but it will increase the time to reach CV to its target value (Figure 13.1).

Some CVs are more important than others from a process point of view. Errors on those important CVs need to be minimized more aggressively. One way to do this is to reduce CV give-up for those important CVs. As shown in Figure 13.2, CV2 is more important than CV1.

It is important to understand that nothing is free: There is always a trade-off between conflicting objectives:

- Control engineers can minimize CV error by reducing the CV give up. This is done at the expense of more error on the other CV's and more movement for the MV's
- Control engineer can minimize MV movement by increasing the MV weights.This is done at the expense of more error on the CV's and more movement for the other MV's

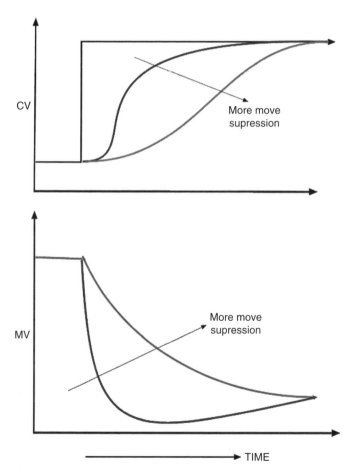

Figure 13.1 Effect of move suppression (or MV weight) on CV and MV trajectory

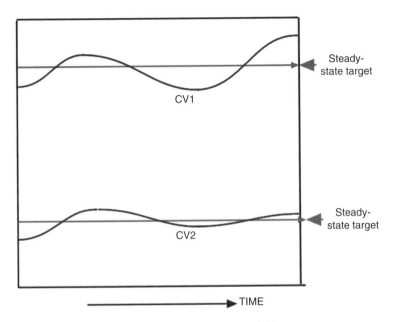

Figure 13.2 Effect of CV give up on CV trajectory and CV error

The basic job for the control engineer is to decide where the inherent variability of the system should go. Control engineers cannot decrease this variability, but control engineers can shift it between MVs and CVs.

The CV give-up should be chosen after discussing with plant production engineer or process engineer.

As MPC is a multivariable controller, any change of one MV weight will affect many CV and other MV movements. Increased MV movement weight on a given MV will have the following effect:

- Moves calculated for the given MV will be smaller.
- Moves calculated for the other MVs may be slightly larger than previously calculated, but this might not have any major impact.
- The ability to eliminate error across the time horizon for the full dependent variable set will be reduced.

MV movement weight will also affect the CV error. The amount of MV movement weight required to obtain the desired MV performance will depend on the dynamic behavior of the CV and on CV give-up. So, the balancing act to determine the exact MV weights is often by trial and error.

Using MV weight to limit the size of the moves has the advantage that normal errors produce normal moves, but large errors produce (and require) larger moves; the move size remains proportional to error.

If in simulation, it is found that a particular MV is continually moving more than the maximum acceptable amount, as specified by the operator, increase its MV movement weight.

If some specific CVs control performance is too slow, then decrease MV movement weights on some relevant MVs where more MV movement weight is tolerable.

Note that there is a trade-off between CV response time, MV movement, and model accuracy. Faster response of CV results in larger MV movement and requires a more accurate model for stable control. On the other hand, slower CV response results in smaller MV movement and works well with a less accurate model. If the model is not accurate, bigger MV movement will cause some impact on other CVs, which the controller could not predict accurately. So those CVs will drift away from their limits if the MV movement is too big, which, consequently, could lead to unstable control. So, the complicacy of model accuracy will also come into play but we would tackle it at online tuning.

Once MV movement weight has been established, the dynamic performance of the explicit CVs should be checked using simulation trends:

- If control of a particular CV is not good, decrease CV give-up for that variable (increase its importance).
- If one or more CVs are being controlled too "tightly," increase their respective CV give-up (i.e., decrease their importance).

With these guidelines, tuning parameters are modified by trial and error in online controller until an acceptable optimal dynamic performance is achieved.

References

Al-Ghazzawi, A., Ali, E., Nouh, A., & Zafiriou, E. (2001). On-line tuning strategy for model predictive controllers. *Journal of Process Control*, 11(3): 265–284.

Ali, E., & Al-Ghazzawi, A. (2003). On-line tuning of model predictive controllers using fuzzy logic. *Canadian journal of chemical engineering*, 81(5): 1041–1051.

Dougherty, D., & Cooper, D. J. (2003). Tuning guidelines of a dynamic matrix controller for integrating (non-self-regulating) processes. *Industrial & Engineering Chemistry Research*, 42(8): 1739–1752.

Garriga, J. L., & Soroush, M. (2010). Model predictive control tuning methods: A review. *Industrial & Engineering Chemistry Research*, 49(8): 3505–3515.

Shridhar, R., & Cooper, D. J. (1997, June). Selection of the move suppression coefficients in tuning dynamic matrix control. In American Control Conference, 1997. Proceedings of the 1997 (Vol. 1, pp. 729–733). IEEE.

Shridhar, R., & Cooper, D. J. (1997). A novel tuning strategy for multivariable model predictive control. *ISA transactions*, 36(4): 273–280.

14

Why Do Some MPC Applications Fail?

14.1　What Went Wrong?

The oil, gas, and chemical manufacturing industries have invested large amounts of money into model predictive control over the years to gain profit by optimizing the operations. In most cases, MPC brings a large amount of benefits to process plants after it is implemented and run successfully for initial years. Within a few years, these control applications often lose their ability to provide optimal profitability due to a variety of reasons. These include process modifications, changes in raw material and product specifications, the commissioning of new downstream units, the degradation of regulatory controls, a lack of training, insufficient monitoring, and a business detachment among planning, engineering, and operations. This chapter details of reasons and investigate what cause the diminishing returns of the MPC benefits.

There are several instances all over world that MPC has provided huge profits for one to two years of its implementation and then profits started decreasing. Some oil refineries also reported that after only four to five years, MPC application did not generate any additional benefits over simple regularity control. So industries are paying attention not only to initial efficient MPC implementation but also to monitoring and support at postimplementation stages to sustain original benefits.

MPC technology has the potential to bring huge profits with a payback within one year. However, the ground reality is that many sites face difficulty in sustaining the profit in long run. Problems faces by these sites are many: their inferential controls does not match with lab, constraints holding schemes are not trusted by operators, and real-time optimizers do not drive the unit to optimal operation. Whenever operators lose trust of MPC application, they keep it offline, and thus the benefit of MPC application diminishes exponentially.

The most common questions at that time are:

- What went wrong?
- What can we do to prevent it from going wrong?
- How we can develop a value program to sustain the MPC benefit over long run?

It has been reported that in large refinery complexes, typically 15 to 20 percent of the MPC installations have failed. Failure is defined as complete disuse of an MPC application. Many companies who installed MPC in the 1990s now realize the need to focus on performance monitoring and continuous support of these applications, rather than only on the implementation of new projects. Figure 14.1 shows the benefit

Multivariable Predictive Control: Applications in Industry, First Edition. Sandip Kumar Lahiri.
© 2017 John Wiley & Sons Ltd. Published 2017 by John Wiley & Sons Ltd.

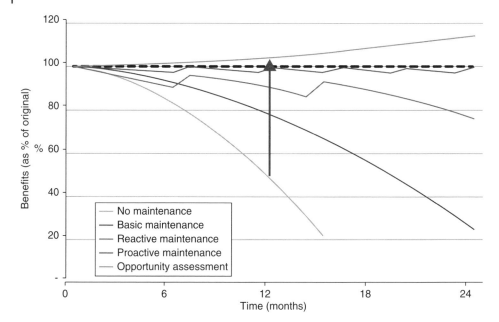

Figure 14.1 Benefit loss over time

Table 14.1 Retaining Initial MPC Benefits after 12 Months

Activity	% Loss of the Original Benefit
No maintenance	45–50% loss
Basic maintenance	15–20% loss
Reactive maintenance	10% loss
Proactive maintenance	5% loss
Opportunity assessment	+2–5% gain

loss over time (trends are indicative only). Table 14.1 summarizes how much MPC succeeded in retaining original benefits after 12 months with or without maintenance. In some companies this reached a critical stage when management began to question the value of new MPC applications after they saw previous ones in their operating plants providing little benefit, sometimes accompanied by negative comments from console operators.

The failure of MPC on the shop floor can be divided into two parts:

1) Failure to build efficient MPC application when it was first build
2) Failure to monitor and protect the original benefits one to three years after initial implementations

It is important to understand this two-failure mode of MPC, their implications, and their root causes. Only then can a proper safeguard be built to minimize the probability of failure.

The following sections describe in detail the reasons behind these two failure modes and how a safeguard can be build.

14.2 Failure to Build Efficient MPC Application

14.2.1 Historical Perspective

The critical question: Where have these sites gone wrong? Many MPC experts (Friedman 1992; 1995; 1998; 1999; Al-Ghazzawi 2003) have investigated the reasons and come up with following explanations. They look to historical development of MPC application and try to find an answer. In the late 1970s and early 1980s, several oil companies (e.g., Shell Oil Company) start developing MPC technology and implementing it in their own sites. The main MPC algorithm development of today's companies was done at that time. Being at nascent stage, oil companies experimented with MPC technology to improve the algorithms and customize them to their own needs. Oil companies developed their own logic control to handle abnormal situations like flooding and cavitation.

Being at the technology development stage, implementing MPC in those days was labor intensive and required multidisciplinary knowledge and expertise. Oil companies who dedicated that level of resource were rewarded by reliable advance control and huge benefit. Then in the late 1980s and 1990s came MPC vendors with generic MPC packages. Decades of MPC technology development came to a halt and MPC implementation shifted to a few vendors such as Aspentech and Honeywell. MPC vendors have limited resources to further develop the MPC technology, but they smoothed the implementation steps by making it as a generic package with all accessories and toolboxes. With packaging, MPC vendors were able to implement MPC applications in refinery, petrochemical and chemical plants across the globe. As that shift took place, competition among the MPC vendors steadily rose, and MPC vendors could no longer afford to implement custom logic as part of MPC project (Friedman 1992). As of now, most MPC applications were built and implemented by MPC vendors. So their expertise, reliability of their adopted methodology, and capability of their MPC software come into play for calculating APC benefits.

Apart from the historical perspective, there are several other factors for failure of MPC applications that can be explained by the following MPC benefit formula.

$$\text{Benefits} = \begin{pmatrix} \text{Optimum} - \text{Current} \\ \text{operation} \end{pmatrix} \begin{bmatrix} \text{Capabilty of} \\ \text{Technology} \\ \text{to Capture} \\ \text{Benefits \%} \end{bmatrix} \begin{bmatrix} \text{Expertise of} \\ \text{Implementation} \\ \text{Team} \\ \% \end{bmatrix} \begin{bmatrix} \text{Reliabilty of APC} \\ \text{Project} \\ \text{Methodology} \\ \% \end{bmatrix}$$

$$[14.1]$$

It is important to understand the contribution of each term to overall failure of MPC. Following section explains that.

14.2.2 Capability of MPC Software to Capture Benefits

The basic algorithms of MPC software of most leading vendors are almost the same. They use a data-driven model to predict future CV values and make changes in MVs to drive the process in optimal fashion. However, the strategy to implement the dynamic control to impose a robust control is quite different among MPC vendors. Each vendor software has some positive and negative points, and no one claims superiority over another.

Five inherent limitations of MPC software are given below:

1) In most cases, soft sensors were built from available lab data. MPC was implemented by taking soft sensor as a controlled variable (CV). Over time, the accuracy of calculation of soft sensor decreases, and MPC takes wrong action based on that.
2) Implementation still depends on the person who implements it. There is no uniformity in strategy of implementation. Knowledge sharing or good publication of MPC implementation is not available. Most of the published papers are marketing types rather than technical.
3) Model development is still done on date from step testing, which is time consuming and prone to error if not monitored properly. The developed model from step test data is as good or as bad as the quality of the data. Step test is not possible for all feed types or for all possible plant capacities. Once the plant is running at different feed composition or deviates largely in capacity, the applicability of the data-driven model becomes questionable.
4) Closed-loop PRBS testing is now possible by most vendors, but their applicability and acceptance at the ground level is still a question mark.
5) Advanced controls make use of steady state and dynamic models. These models rely on many instrument readings. If any of the measurements are erroneous, the model and associated control actions are no longer valid. Furthermore, advanced controls push the plant to constraints, and that makes operators uncomfortable. They demand (rightfully) that at constraints, control schemes perform perfectly. They (rightfully again) turn off control schemes upon imperfect performance. We must conclude that in spite of improved packaging, robust control applications necessitate "real time HAZOP" logic. Lack of HAZOP analysis, and subsequently no trouble avoidance logic makes the MPC application ineffective.

14.2.3 Expertise of Implementation Team

Effectiveness of MPC application still depends on the expertise of the implementation team, which consists of MPC vendor team and client MPC implementation team. In most cases, for a new project, the MPC vendor team is unaware of the chemical process and its details. On the other hand, the client plant team has poor knowledge of the various features of MPC software, its working principles, and so on. The key to successful implementation of MPC applications is how much these two different background teams communicate, understand, and share knowledge. These two teams have different backgrounds and knowledge base. How they amalgamate their knowledge with each other—that defines the success.

14.2.3.1   MPC Vendor Limitations

MPC vendors are much in demand. They are busy trying to implement new projects one after another within the shortest possible time. That contributes to four major limitations:

1) *Vendor has no time to monitor the efficiency of implemented MPC:* MPC vendors are few and overloaded with contracts. They don't have time to monitor and maintain implemented MPC applications. Sometimes they take very little time for online tuning and ignore any deficiency of the application. Once they finish implementation, they hand over the whole application to client plant personnel and fly away to another project. In most cases, client plant personnel are not well equipped to tackle the issues. Control engineers, who do not have time for analysis, respond to problems by detuning the controls. They keep service factors high by weakening incorrect controls to the point of being harmless.

2) *Poor understanding of the process:* The prerequisite to develop good MPC application is to understand the process. It needs time to do thorough readings and analysis of process data to develop a detailed understanding of the process. There is no shortcut. Once the process and the constraints of the plants are fully understood, then an effective MPC application can be develop that will exploit the margin in the process. In most cases, however, MPC vendors heavily rely on the client plant process and production engineers' opinions and try to develop the MPC application without thoroughly understanding the process. The net result: Developed MPC applications are able to control the process (in most cases) but fail to derive any benefit from the process by exploiting the margins. This is one of the major reasons for failure of MPC application when it is first build.

3) *Limited time to understand the reactor:* Reactors in commercial plants are less explored equipment as far as MPC is concerned. Usually, reactors are the equipment that drives the economics of the process. For catalytic reactors, selectivity or yield improvement has dramatic effects on process. But as the knowledge of industrial reactors are poor, in most cases reactors are running within the defined boundaries of process parameters dictated by the technology licensor. Efforts to optimize the reactor conditions are minimal. The job of a good MPC application is to study the effect of different process parameters on reactor performance (selectivity, yield, and throughput) and make a control system that implements such optimization in a reactor loop. But, as there is a very short time available to MPC vendors, usually they ignore any optimization scope in reactors and try to build a control system that just controls the reactor parameters rather than optimizing them. This leads to huge loss of potential benefits that could otherwise be tapped through MPC.

4) *Nonavailability of strategies or published papers for MPC application:* There are few good technical papers and book that discuss the different strategies of MPC application. Most of the available papers are the advertisement or marketing type of the MPC vendors, with little technical values. Also, there are no international standard for functional design as there are for API, ASME etc. Due to this limited knowledge sharing, MPC application has not developed as fast as it should have.

14.2.3.2 Client Limitations

Most engineers from client plants have limited knowledge of MPC technology and its different features. Hence, they cannot get the full picture of how MPC will exploit the margins and fetch benefit.

As stated earlier, prerequisite to building good MPC application is to amalgamate the knowledge of plant engineers (who have domain knowledge) and MPC control engineers (with MPC knowledge). As they don't know each other's field completely, in most cases they cannot engage in a meaningful knowledge sharing that will promote good MPC implementation. Some of the factors that contribute to poor communications between plant engineers and MPC engineers are as follows:

- No one in the plant understands all the functions of process engineering, plant operation, and control engineering.
- The plant engineer was a given this MPC responsibility on top of routine job responsibilities, and is working in MPC implementation as time permits. This limits the engineer's involvement and performance considerably.
- Plant engineers were given MPC training at the end of the project instead of at its start. Hence, throughout the project execution time, the engineer remains blind as to what is going on.
- In most cases, plant engineers don't have a good idea about the limits or constraints of the process.

14.2.4 Reliability of APC Project Methodology

There is no standard for MPC project methodology and how to do a functional design in step-by-step method. Each vendor and each MPC team apply their own knowledge and experience to develop a good MPC application. As there is no standard methodology, the probability of failure is high. It pays to ask hard questions at the functional design stage of an advanced control project. These questions reduce the agony of discovering operational problems by trial and error. It is seen in various refineries and petrochemical plants that they were in the habit of coming up with scheme designs, and implementing and commissioning them, only to find out that these schemes were problematic and often hazardous (Friedman 1992, 1995, 1999). They would then modify the design, reimplement it, and recommission it to find the schemes lacking again. They might go through three or four cycles and end up with an "afterthought" instead of a proper robust control application.

MPC expert (Friedman 1992, 1995, 1999) suggested that a committee should be formed to review operability questions. Review meetings should be conducted twice: first to review the functional design and second just before a scheme is commissioned. The first meeting should consider the plant economic objective and confirm that it can be reached with the proposed design. The second meeting is for discussion of specific constraint limits, actions to be taken when encountering unusual disturbances, and what precautions should be taken during testing of the new scheme.

It is important to spend much time analyzing what could go wrong and install logic to face such situations. Promise operators that control schemes will never act in unsafe manner. Then work to keep your promise.

If the objective of a scheme is maximization or minimization, the committee must understand the constraints for that piece of equipment (Friedman 1992, 1995, 1999). There is often a need to discriminate between a "constraint" and a "myth."

An example of a constraint is: If you go beyond a certain limit, a distillation column will flood. An example of a myth is: "Last time we increased this flow, the unit went unstable." There may be a real constraint behind the myth, but it needs to be understood and explained in engineering terms (Friedman 1992, 1995, 1999). Sometimes operation people take the PFD values as the limit or constraints to operate the equipment. Normally, licensors, designers, and equipment vendors keep a design and safety margin, respectively, while designing the equipment. So the installed equipment has all three margins usually available. These margins or constraints need to be evaluated from the actual operation and cannot be simply determine from the PFD figures or equipment data sheet figures.

As for example, one refinery had a feed maximization control scheme where one of the constraints was the "nominal" throughput of that unit. The control scheme brought the feed flow to its nominal throughput and then stopped. No equipment was operated to a limit, and while the scheme was kept "on" with 100 percent service factor, it was not making money for the refinery (Friedman 1992, 1995, 1999).

Having understood the constraints and disturbances, it is necessary to look at the control scheme holistically. More often, it was seen that tighter control of one piece of equipment affects the other unit control in downstream section negatively. Control performance improvement of one piece of equipment at the expense of other equipment is not desirable; a holistic view of overall control is necessary in such cases.

Analyzers in advanced control schemes are often the bad actor in a control loop. Before ordering an analyzer, verify that it is simple and maintainable; that the sample point and sample loop are such that the dead time will still be reasonable; and that it is the simplest analyzer to do the job. If the analyzer is already installed, study the historical repair record and how well the analyzer agrees with the laboratory.

Finally, one should consider the consequences of erroneous input measurements into the scheme and how to protect the unit from such consequences. During the operability review meeting, decide which measurements are critical to the scheme and what tests are required for the specific measurements (Friedman 1992, 1995, 1999).

14.3 Contributing Failure Factors of Postimplementation MPC Application

Why do service factors and MPC performance declines over time? Finding the answer to this critical question is key to success.

Low service factors and MPC performance deterioration over time can be related to the equipment, to software, or even to human aspects.

Many MPC experts (Friedman 1992, 1995, 1998, 1999; Al-Ghazzawi 2003) have investigated the possible failure factor of MPC application. Their findings include (but not limited to) eight factors that contribute to poor performance or failure of installed MPC applications after implementation. These are shown in Figure 14.2 and described in the following sections.

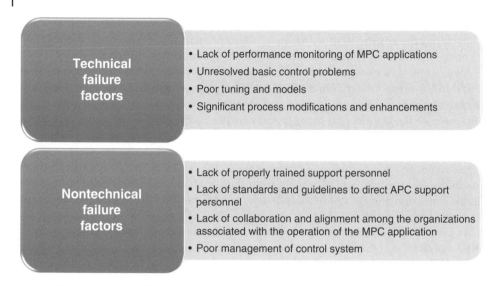

Figure 14.2 Contributing failure factors of postimplementation of MPC applications

14.3.1 Technical Failure Factors

14.3.1.1 Lack of Performance Monitoring of MPC Application

Most of the plants don't develop performance KPIs to monitor the controller performance:

- *KPIs are not being calculated:* Various KPIs representing the performance of the controller are not developed and most of the time not calculated online in real time. As these performance KPIs are not available, operators and engineers remain unaware when controller performance degrades.
- *No model quality KPIs related:* A good model is the building block of any MPCA. Model quality KPIs need to be developed and monitored periodically to track the model applicability and goodness in changing environment over long run. Verification of model quality can be done by measuring unbiased predictions and prediction errors. Low prediction errors all the time at least for key variables will give an idea of how well the models are built.

As the focus of the MPC engineer turns more and more from project implementation to controller maintenance, a monitoring and diagnostic tool is required to effectively check the health of MPC applications and to provide troubleshooting capabilities. A monitoring tool is essential to check MPC status and diagnose problems in a timely fashion. Remediation efforts can then be taken to keep the MPC applications online with optimal results.

14.3.1.2 Unresolved Basic Control Problems

The base or regulatory control layer consists of the field sensors and final control elements, such as valves. This layer includes the basic control on the distributed control system, (DCS), such as the proportional, integral and derivative (PID) controller. It may also encompass other forms of advanced regulatory control, such as cascade, ratio, and constraint control. Typical problems encountered are as follows:

- *Instrumentation problems:* MPC acts as supervisory control and it implements its optimization capability through the base-level regulatory control. Common problems in this area include faulty instrumentation, nonfunctional analyzers, valve stiction and hysteresis, and unsatisfactory performance of the DCS-based regulatory controls.
- *Computer failure (servers):* Servers failure connected with MPC controllers is another common problem.
- *Network failure:* Data collection network failure is very common.

14.3.1.3 Poor Tuning and Degraded Model Quality

- *Poorly tuned controllers:* A multivariable predictive control application typically contains dozens of tuning parameters that may require periodic adjustments. Inappropriate choices of weighting factor, priorities, move suppressions, step sizes, and economic cost factors will cause controller problems. Online tuning also should be done periodically, as the process characteristics unmeasured disturbance changes over time.
- *Inadequate process models:* Good model quality is critical to optimal performance. Poor model quality may be the result of nonlinearity not taken into account, poor initial identification of models, unmeasured disturbances, and process changes as discussed earlier. Model problems may also be the result of including models that should not be included (due to very small gains) or vice versa, or not including models that should be included.
- *Software problems:* Sometimes software are not compatible with other components of control systems. In some incidence, software takes too much time to extract data from DCS and write back set point to DCS due to data overload or memory limitations. These problems severely impact the controller performance.

14.3.1.4 Problems Related to Controller Design

Control matrix is more complex than it should: Sometimes insignificant MV-CV relationship in model matrix can severely deteriorate controller performance. Simple model matrix, which captures the process dynamics, is the best matrix.

- *Problems with lab analysis time stamps:* Soft sensor performance can be monitored by plotting lab analysis and soft sensor indication side by side. It is important to maintain accurate time stamp of the collected samples, as the soft sensor will take the relevant process parameters at that point of time to generate the soft quality parameter value.
- *Problems with lab analysis feedback:* There should be a good procedure to input lab analysis data into the online soft sensor software so that it can make a bias correction.
- *System not designed for product transitions:* In most cases, step test was not done during polymer product transition time. So controller performance badly impacted during product transition time as the model curves are not applicable during that time.

14.3.1.5 Significant Process Modifications and Enhancement

In a competitive environment, it is assumed that continuous process modifications and enhancements are required to meet product demands and corporate profitability. These modifications generally increase model and inferential error by impacting the dynamic models that were initially developed from actual plant testing for the MPC. These modifications and enhancements come from a variety of sources, including different feedstock material, changing business targets and product specifications, new equipment, changes in upstream or downstream operations, and de-bottlenecking projects. These items may result in new limits and constraints that, if not properly addressed, reduce the ability of the application to effectively control and optimize the plant. The entire design of the MPC application may need to be reconsidered depending on the modifications performed.

14.3.2   Nontechnical Failure Factors

14.3.2.1   Lack of Properly Trained Personnel

This includes a deficiency of dedicated and competent process control engineers to support, monitor, and maintain installed MPC applications. This deficiency naturally leads to poorly trained operations and other engineering staff, since the process control engineer is the main trainer of MPC technologies within a facility. Statistics compiled by Solomon & Associates show that more than 75 percent of the process industry believes that the lack of trained and experienced control engineers, as well as trained operators, is always a limitation to capturing control long-term benefits (Al-Ghazzawi, Anderson, & Al-Soudani (2003)). Some typical problems are given below:

- Operator lacks the necessary training.
- Operator doesn't know how MPC works.
- Operator doesn't know how to run the application.
- Operator lacks experience with the system.
- Operator lack real-time support.
- Excessive judgment is required from the operator.
- Operator doesn't know when he can turn controllers on.
- Operator is reluctant to turn the controllers on when he doesn't know why they were turned off.
- Operator doesn't trust the system.
- There is a failure to report corrective measures that have been taken.

14.3.2.2   Lack of Standards and Guidelines to MPC Support Personnel

One of the problems identified by large refinery and petrochemical complex in the implementation and support of MPC applications has been the lack of consistency across several projects and post installation support. Best practices of one project should be captured and repeated to other projects. All project steps performed at one plant may not be accomplished at another for various reasons. This has resulted on occasion in missed opportunities for knowledge transfer and confusion on how to best structure support. A clear set of MPC guidelines is needed that cover areas of project execution procedures, pre-audit and project justifications, post-audits, operator guidelines, plant control engineer guidelines and performance monitoring.

14.3.2.3   Lack of Organizational Collaboration and Alignment

Successful MPC implementation requires a team effort by members representing the operations, maintenance and engineering organizations of a process facility. Key success factor of MPC projects depends on the amalgamation and synergy of these diverse team. It is important that each team member know his responsibilities as related to MPC operation along with expectations of other team members. Any omission of cohesiveness among the organizations usually results in a loss of business direction.

14.3.2.4 Poor Management of Control System

MPC should be considered as a system (like ERP) and all the standard operating procedure applicable to system in plant should be applied to MPC. Common problems encountered in this area are as follows:

- *Management of change:* Any change in MPC functional design, different limits, and tuning parameters should be well studied, challenged, and properly documented and pass through management of change procedures of the plant.
- *Poor knowledge transfer from vendor to client engineers:* Challenges encountered during the implementation and support of some of these applications were related to the project execution methodology in which vendors were in complete control of all project phases, with minimal participation from company engineers. This resulted in very little, if any, hands-on experience and technology transfer.
- *Process modifications:* Any process modifications that can impact MPC performance should be well studied and required change in MPC design should be performed.
- *New grades/recipes require model updating:* MPC functional design should be a dynamic document and the model should be updated for new grade of product or feed composition change.
- *Staffing changes causing loss of knowledge:* Plants should develop a training procedure to capture and retain the knowledge base of MPC. Normally, change of knowledgeable staff has a negative impact on MPC understanding of the site.
- *Lack of visibility of the application benefits:* MPC benefits should be clearly quantified, and related KPIs should be developed so that all interested parties have a clear understanding of the profit MPC brings.
- *Results are poorly reported:* MPC performance, profit gain, MPC problem, and maintenance issues should be clearly reported to all the stakeholders.
- *KPI calculations are complex and time consuming:* KPI should be made simple and online automatic calculations should be developed.
- *Lack of commitment:* Other obstacles were related to a lack of appreciation and understanding of MPC from the various entities within the operating facility, and therefore, absence of a dedicated team within the plant to coordinate and facilitate project activities.

    MPC requires commitment from different discipline and departments. It should not be considered as operation department responsibilities.

    Problems are not efficiently diagnosed and reported to the MPC engineers: Any MPC problem should be studied in detail and properly reported to concern MPC engineer.
- *Corrective actions take too long to be implemented:* Sometimes, MPC requires adding a model or do a step test to get a new MV-CV relationship. In most cases, it takes a long time to do such modifications. Limited staffing is dedicated to maintenance: MPC requires a sufficient number of dedicated control engineers for maintenance so that its performance can be sustained.

14.4 Strategies to Avoid MPC Failures

Most of the large sites that implemented MPC have experienced one or more of the above problems on various MPC installations, which eventually led to their complete disuse.

Industries where MPC performance degrades over time are asking the following questions:

- What can we do to improve advanced control and optimization performance?
- Are we doomed to fail, or is there a way to make the technology work?
- How can sites recover from operator distrust syndrome?

What are the long-term measures to prevent such failures and sustain the benefit? There is threefold motivation behind this renewed support focus:

1) Safeguard the original MPC investment
2) Capture the original and full economic benefits that are usually sustainable from year to year and generally have a rate of return many times over the original investment
3) Capture the intangible benefits that are oftentimes difficult to quantify, such as increased process stability, greater operational reliability, increased safety, and reduced equipment wear

As already discussed, there are some technical and management issues that are responsible for performance degradation of MPC. The bad actors and technical issues should deal with technical competency. To deal with management issues effectively, proper systems and procedures should be developed. Some ways to avoid future failures are summarized in Figure 14.3.

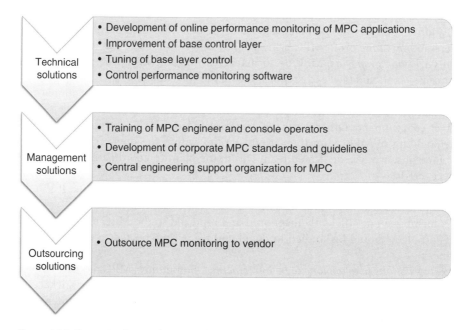

Technical solutions
- Development of online performance monitoring of MPC applications
- Improvement of base control layer
- Tuning of base layer control
- Control performance monitoring software

Management solutions
- Training of MPC engineer and console operators
- Development of corporate MPC standards and guidelines
- Central engineering support organization for MPC

Outsourcing solutions
- Outsource MPC monitoring to vendor

Figure 14.3 Strategies for avoiding MPC failures

14.4.1 Technical Solutions

14.4.1.1 Development of Online Performance Monitoring of APC Applications

One problem that has hampered progress in process control throughout the industry is lack of monitoring how the advanced control schemes perform once they are commissioned. In most sites, people did not know the service factors of schemes, did not have any measure of the performance of the advanced control schemes when they are working, and did not know whether the advanced controls actually perform better than no controls. There were endless arguments about when a scheme has been turned off and why, and whether a scheme is really accomplishing its objectives.

It is only through objectively monitoring the performance of control schemes that one can learn how well a scheme is performing and when it is need of maintenance.

An organization's employees can never be sure of the success and effectiveness of its MPC program without monitoring the performance of MPC installations. Appropriate measures or key performance indicators (KPIs) must be defined, calculated from validated data and archived for retrieval by MPC engineers that are responsible for ensuring that the MPC applications are optimal. Earlier chapter gives an overview of an MPC performance-monitoring KPI that is based on internally defined metrics that provide high-level monitoring of multivariable controllers.

The ideal single measurement for MPC performance would be a comprehensive KPI that shows the economic performance of the MPC. This would be the only necessary metric. However, in reality, due to rapid dynamics of a real process and industrial economics, any such economic measurement would only be very approximate and, oftentimes, a misleading indicator of the performance and value of the MPC. Therefore, several KPIs are required and, when used together, can be interpreted to give a reasonable picture of MPC performance.

Some practical guidelines are given below:

- *Make sure KPIs are defined:* Common KPIs that need to be monitored are given in earlier chapters. It is good practice to layer these KPIs in at least two layers, defined as a high-level layer and a lower tier. They should be simple and portray concise information. MPC is a complex control technology that is composed of many components and subcomponents. Monitoring such technology can be complex as well. The challenge is to identify KPIs that are of a high enough level to indicate controller degradation without confusion and without overburdening MPC support engineers. They should provide immediate key statistics with the ability of the engineer to solicit additional information through lower-level KPIs on-demand through trends and "drill-down" processes if the higher-level indices indicate a problem. The additional drill-down information and historical data could well be provided by either the original MPC product vendor or a third-party application.
- *Automate KPI calculations:* Online calculations and trends to be developed for those KPIs so that every stakeholder can see it from their desktop. The high-level KPIs should be readily available at the engineer's desktop through web-enabled technology.
- *Use KPI in decision-making process:* The economics KPI should be used to decide when to plan for a preventive or reactive maintenance. Those KPIs should be linked to maintenance plan, who will act and how if the performance KPIs touches a lower limit. These KPIs should be used to highlight the amount of loss incurred by the plant due to nonperforming of MPC and to justify investment in maintenance of MPC.

14.4.1.2 Improvement of Base Control Layer

MPC cannot work efficiently if base control layer or regulatory control layer is weak. Deal first with DCS problems. Do not attempt advanced control on top of ill-configured or poorly tuned basic controls. Hence, strengthening base control layer is an important prerequisite to build the good foundations. As discussed in earlier chapters, reasons for poor base layer are many, such as sticky control valves, improper valve sizing, faulty positioner, improper measuring transmitters such as flow and pressure transmitters, and improper controller tuning. All these need to be identified and rectified before we jump to implement MPC. Another common bad actor for MPC is bad analyzers. In most cases, product impurities are considered as an important CV in MPC, and any failure to obtain reliable measurement of this CV impacts the performance of MPC very badly. Online analyzers are mostly used to analyze product impurities in ppm or percent level, and their measurement is absolutely necessary for MPC, so that it can push the plant until product spec are not violated. Study shows that on average, 40 percent of the benefits of MPC rely on our ability to measure product qualities online. Some analyzers can be replaced by inferential controls, though the latter are not always available and not on every unit. State-of-the-art models can capture perhaps half the product quality control benefits without analyzers. Still, at least 20 percent of advanced control benefits depend on analyzers being properly specified, installed and maintained.

But the problem with the online sophisticated analyzers are that they are very sensitive and need proper periodical maintenance to keep them online. The ground reality in most sites is that plants spend the money to buy analyzers but do not allocate the manpower to support them.

What can be done, then, to improve benefits associated with quality measurements? The following guideline should help (Friedman 1992, 1995, 1999):

- Only economically justified analyzers should be purchased.
- Analyzers whose mechanism deviate significantly from laboratory procedures should be treated with suspicion, unless there is a body of evidence to support good correlation between analyzer and lab.
- Avoid analyzers that are a known maintenance headache, such as distillation analyzers working above 700 °F or optical analyzers, which require frequent cleaning.
- Sometimes product specifications call for difficult analyzers measuring impurities at the ppm level. Often, in such units, though, there is another stream much easier to measure whose quality trends with the ppm level product quality. It pays to consider these easier, simpler options rather than blindly specifying analyzers on the final products.
- Take a close look at the sample loop designs. Analyzers frequently fail because of dirty, moist, or corrosive sample systems, or sometimes simply because of an inadequate fast sample pump or insufficient pressure drop across the fast loop.

14.4.1.3 Tuning Basic Controls

Building advanced controls on poorly tuned basic controls is like building skyscrapers on quicksand. But, in most plants, this is the reality. The art of loop-tuning demands a combination of skills not often found in one individual: process dynamics and control theory; process engineering and understanding of inter-loop interactions; and familiarity with the economic driving forces and operating objectives. The skill requirements, and also the fact that tuning is a time-consuming activity, lead to tuning practices such as cutting or adding to the gain whenever someone complains, without conducting a test to identify the loop dynamics and without appreciation of the tuning objective for the loop.

The most commonly mistuned loops in the audited refineries were level controllers, where the operating objective is usually to keep the flow steady and accept level swings, but where tuners often prefer the opposite. At best, poorly tuned level controllers will introduce unnecessary disturbances in downstream equipment. At worst, the flow swings will interact through heat exchangers and recycles to drive the whole unit unstable without anyone knowing how to discriminate between cause and effect. The tuner would then respond by detuning the wrong loops adding sluggishness to the instability.

Some practical guidelines are given below:

- Identify the people permitted to tune loops and train them well.
- Improve communication between the instrument technician who tunes the basic control loops and the process control engineer who tunes the advanced controls.
- Keep a history of tuning changes and reasons for changes. Hopefully, this will eliminate perpetual tuning and detuning of the same loops.
- Create a guideline for tuning to cover the most common tuning objectives for flow, pressure, temperature and level loops.
- Create a standard for predictive control for analyzer loops with excessive dead time.
- Do not shy away from spending time on tuning loops. An expert may be able to tune 10 loops per day with the usual refinery mixture of loops.
- Advanced control loops are slower and even more time-consuming.

14.4.1.4   Control Performance Monitoring Software

As a result of the industry requirements for MPC monitoring tools, several leading process automation technology providers have developed tools to help monitor the condition of MPC applications. These systems address the needs of users to ensure that these controllers are functioning properly to ensure tangible benefits are realized from their operation. Examples of commercially available MPC monitoring tools are presented in Table 14.2.

These tools provide MPC support staff with comprehensive information on controllers' performance, effectiveness, model accuracy, as well as a range of performance benchmarks for both the MPC controller and the regulatory or PID control layer. Such tools have contributed greatly over the past few years in ensuring that installed controllers are functioning properly for prolonged periods.

Table 14.2 Commercial MPC Monitoring Tools

Company	Tool	Supported MPC Technology	Website
Aspen Tech	Aspen Watch	DMCplus	www.aspentech.com
Honeywell	APC Scout	RMPCT	www.hps.honeywell.com
Shell- Yokogawa	MDpro	SMOC	www.yokogawa.com
Matrikon	Process Doctor	RMPCT, DMCplus	www.matrikon.com

Opting for control-performance-monitoring software makes the job of sustaining MPC assets much easier. Today's web-based software augments human involvement by automating the collection of relevant data and distributing these data in a more meaningful and effective way. Condition-based applications are particularly beneficial for analyzing control asset performance. They can be used to configure metrics and reports that pinpoint the automation and process controls not operating to their potential, as well as provide detailed analysis on the root cause of the performance degradation. The ability to do this from historical data ensures a plant can benchmark against its own goals or other facilities without the need for comprehensive testing.

14.4.2   Management Solutions

Investments in process control applications have to be accompanied by investments in process control users and staff to reap and sustain benefits from applications. It is very important to create a continuous training system for MPC Engineer Development Program to develop adequate in-house skills and resources to properly implement, support, and maintain MPC applications to achieve sustained benefits from installed applications.

14.4.2.1   Training of MPC Console Operators

Operators are the drivers of MPC scheme. No MPC scheme will be successful without the blessings of operators. Each board operator needs to understand the objectives of advanced control schemes, constraints, overrides, and logical structure. It is a good idea to spend about 30 minutes per shift per control scheme at commissioning time plus repeated sessions if a scheme is modified. Operators should be involved at the early stage of functional design, offline simulation, and online tuning so that they can follow what is going on. This should be considered as an essential requirement to sustain MPC benefit. Ground reality is this: Engineers spoke down to operators, did not take the time to make sure each operator understood, did not solicit ideas from operators, and by and large did not accept that their mission is to build tools in service of the operators.

MPC sometimes took actions that are not as usual as operator used to take. That's why MPC is superior to operator. It is essential to train the operators regarding why MPC is taking the actions in that particular fashion. What future constraints it has predicted to hit and how it is moving the process not to cross the future limit. It is not easy task, but operators must understand how MPC works. Only then will operators have faith in the MPC system.

Operators have the final responsibility for safety of the unit, and to that extent they must be permitted to turn control schemes off whenever they perceive a safety or operability problem with the scheme. On the other hand, when the operator turns off a scheme he or she should help by filling out a trouble report form explaining why the scheme was not performing and supporting the report with plots and other data. Once a trouble report is out, it should be treated seriously and some corrective action must be taken. That action may simply be a change of display, more operator training, additional constraint in the scheme or sometimes a more drastic change of logic before the scheme can be re-commissioned.

14.4.2.2 Training of MPC Control Engineers

It is important to include MPC requirements in the career development plan of panel operators and control engineers. The training program objectives are:

1) Develop competent and self-reliant in-house MPC engineers capable of supporting and maintaining existing applications, in addition to identifying future MPC applications opportunities at their facilities.
2) Implement MPC applications at various company facilities as "hands-on" training for program participants.
3) Ensure that MPC application benefits are acquired and continuously sustained within the company.

The MPC Engineers Development Program is structured to provide a combination of both classroom training and "hands-on" experience. The major emphasis is on gaining field experience from active participation in the implementation and support of MPC applications.

It is also necessary to develop a set of MPC engineer competencies to identify the skills an MPC engineer needs to master and attach them with their career progression. MPC engineers are the ones who understand the operating variables and how they interact, and the mechanisms for advanced controls to make money. About 80 percent of advanced control incentives are related to what can be termed *online process engineering* or figuring out how the unit should be operated and driving it toward that objective without violating any constraints.

Once process-engineering skills are established, it is of value to the control engineer to be versed in control theory, and there ought to be at least one individual in each plant who understands both process engineering and control theory.

In most of chemical plants, it was difficult to find a person with both process engineering and process dynamics skills. The process control groups generally were composed of instrument engineers without process engineering skills, process engineers without knowledge of control theory and chemical engineers who were hired from school or from other industries with an understanding of dynamics but insufficient training in actual process. The main aim of MPC engineer development programs is to develop professionals with both process engineering and control mindsets.

Suggested training starts with a basic course in dynamics and tuning of control loops. The course is very general and intended for new process control engineers as well as instrument technicians and operators. Then follows an intermediate-level course to cover the most important advanced control techniques and their application to a process (e.g., furnace). It should also involve hands-on training to build soft sensors. A third course is more advanced, covering the very difficult problem of controlling distillation columns. Given that this is the most common equipment in a refinery/petrochemicals and also the most difficult to control, it is useful to introduce it as early as possible, but not before the participants have had a chance to experience the control of simpler equipment. There are also courses for specific units, such as crude unit, FCC, or reformer. These courses cover a mixture of process engineering considerations and objectives, plus the typical dynamics of the unit and how to control it. These courses do not have a wide audience and are, therefore, expensive. Still, a control engineer without experience on the unit for which he or she is responsible will benefit greatly from taking such a course.

MPC engineers should be involved at every stage of project execution, and there should be a mechanism that facilitates knowledge transfer between MPC vendors and plant MPC engineers. Plant control engineers should know all the steps taken and the jobs done by the vendors, starting from the early functional design stage to final implementation stage. It is important to create a systematic procedure of knowledge management that encourages knowledge sharing.

14.4.2.3 Development of Corporate MPC Standards and Guidelines

It is standard practice in industry that any piece of equipment installed in the chemical plant must conform to some design standards, such as the ASME code, TEMA, or others. However, it is surprisingly observed that the same plant readily accepts control designs that do not conform to any standard. While there is no official book of advanced control standards blessed by API or any reputed organization, there is still a need to create a set of standards, or else there is no guarantee of safety, operability, or maintainability.

In many sites, there were no documentation or design practices that would even remotely resemble a standard way of doing anything.

It is important to establish standard procedures for advanced control designs, interfaces, HAZOP analysis, and real-time protecting tools, engineering and operator documentation, and so on. That improves safety and reduces maintenance in the long run.

To address planning, implementation and support requirements for MPC applications, it is necessary to develop a set of *standard operating procedures* and guidelines to assist plants and central engineering staff in supporting installed applications. These guidelines cover APC routine monitoring and troubleshooting requirements, performance monitoring, post-audit, and evaluation.

The proposed MPC standards and a brief description of their scope and purpose are given below:

- *Guidelines and standards to follow during MPC execution:* This guideline covers the typical execution steps of a multivariable control implementation project. It starts with the preliminary design step and goes through commissioning, as-built documentation, and touches on training. It describes each step in detail and includes deliverables and personnel requirements. It should cover the vendor responsibility and client plant personnel responsibility and deliverables.
- *Guidelines for pre-MPC audits:* The pre-APC audit guideline covers the assessment of the regulatory control layer, as well as identifies the process units where significant benefits may be realized from APC. This guideline provides the methodology to establish the pre-APC base case, as well as an estimate of the tangible APC benefits with an acceptable degree of accuracy. This also gives the statistical methodology used for cost benefit calculations before MPC implementation.
- *Standard operating procedure for routine monitoring and troubleshooting:* Routine APC support tasks are associated with the four periods: daily, weekly, monthly, and annually. The guideline describes in full detail the various tasks required by support engineers during the four periods. It should consists the following (but not be limited to):
  - o Define how reporting and tracking of application issues will be executed.
  - o Standardize performance follow-up reports (format, frequency, distribution).
  - o Implement the communication plan.

- o Define individual and collective goals.
- o Define scope and quality criteria for support contracts.
- o Routine—Some actions must be regular, to avoid gradual service factor decay without stakeholders know it.
- o Verify application status (daily).
- o Issue follow-up reports (monthly).
- o Update visible management panel (monthly).
- o Follow-up meeting, including process, production, operations, automation and maintenance teams (monthly).
- o Follow-up on reported issues (monthly).
- o Prepare/update training material (annual).
- o Train operators, shift supervisors and production engineers (annual).
- o Audit application (annual).
- o Occasional—Report when application presents problems or hardware and software undergo modifications.
- o Create a systematic procedure of knowledge management that encourages knowledge sharing.
- *Post-audits for MPC applications:* Post-audit studies characterize how well an advanced process control (APC) project satisfied its objectives. The post-audit guidelines present different approaches to benefits estimation, as well as several guidelines to ensure a consistent basis of comparison between the before-and-after APC process information. To reap the most benefits from an APC application, it is necessary to monitor and analyze the APC application performance; in evaluating an APC project, it is necessary to evaluate the application's control, process, and economic performance measures. This guideline also emphasizes the need to highlight intangible benefits, along with those that can be clearly quantified.

14.4.2.4 Central Engineering Support Organization for MPC

The proven economic benefits associated with advanced process controls deployment prompted the central engineering organization for large-scale refinery or petrochemical complex to organize an Advanced Process Control Unit. This unit's mission is to provide superior leadership in the application of MPC technologies to capture and sustain profitable opportunities (Al-Ghazzawi 2003).

The unit's vision is to be the recognized center of excellence in the application of MPC technologies within large-scale refinery or petrochemical complex. This unit should provide consulting expertise for the planning, implementation, and support of MPC applications to all company facilities. These engineers are considered second-level support engineers, providing overall support to the Level 1 MPC engineers that reside in the plants. The duties and responsibilities of both first and second level MPC engineers are given below.

Responsibility of plant level MPC engineer (level 1) (Al-Ghazzawi 2003):

- Daily monitor and support MPC applications. Ensure that economic objectives are current.
- Respond to operators' questions, and concerns.
- Provide periodic training for console operators.
- Perform minor modifications to MPC applications. These include tuning and minor model modification.

- Assist in the identification and justification of MPC applications.
- Provide feedback to plant management on performance and value of installed MPC applications.
- Serve as liaison among operations, maintenance, and engineering.

Responsibility of central level MPC engineer (level 2) (Al-Ghazzawi 2003):

- Create service level agreements with plants.
- Support new MPC projects implementation.
- Provijk de contract and procurement support and assistance.
- Provide support for major MPC modifications.
- Coordinate and provide training for level 1 and level 2 MPC engineers.
- Provide higher requirement needs of APC expertise and support.
- Coordinate software upgrades and new releases.
- Track industry trends in MPC technology and evaluate new products.

Other responsibility of central MPC groups are as follows (Al-Ghazzawi 2003):

- Team alignment process: One of the major responsibilities of central engineering support group is to align the MPC team with a common goal. As successful MPC implementation requires commitment from multidisciplinary groups namely process engineering, production engineering, control system engineering, IT department, instrumentation department, planning department, local management, and MPC vendor experts. This is a planning process that helps establish high-performance teams that work effectively to achieve desired results. Through this process, team members work harmoniously toward a common mission and vision. Throughout the team alignment process (TAP), team members identify a number of high-leverage items that will assist in the transition toward the team's vision. Project team members worked together in developing a project mission and vision statement, and they established a set of team norms to ensure team members are all committed and accountable to achieve their common purpose. Define individual and collective goals.
- Build operator-focused tools to support the activities they'll have to execute.
- Define scope and quality criteria for support contracts.
- Install visible management panels in the control rooms.
- Diagnose application problems.
- Adjust base layer controller tuning.
- Adjust MPC controller tuning.
- Add new MVs or CVs to the MPC controllers.
- Update existing models: Upgrade hardware and software.
- Address DCS, PIMS, and LIMS configuration.

Some important advice from industry expert (Friedman 1992, 1995, 1998, 1999) are summarized as follows:

1) Plant management should be made aware that sophisticated MPC technology should be matched against the number and quality of MPC people it is willing to dedicate. What is the point of purchasing sophisticated technology that no one in the house can support?
2) Plant should deploy dedicated MPC persons, and that should be their main focus. Double-dipping does not work.

3) Operators, process engineers, and others should be encouraged to report problems. Listen their concerns. Fix problems quickly and keep operators informed of the nature of problem and method of repair.
4) Set up real-time KPIs for reporting real economics of advanced control. How would one convince management to increase maintenance effort without showing that advanced controls improve plant economics?
5) MPC contractors' work must culminate in a complete technology transfer and detailed documentation.

For sustained economic return, there is still no alternative to good engineering, dedicated maintenance, and lots of attention to details.

14.4.3   Outsourcing Solutions

Nowadays, APC vendors come up with special services to monitor the MPC performance across the plant. Customers can completely outsource the MPC performance monitoring to these MPC vendors. These service contracts deliver value to the customer by providing diligent monitoring of performance metrics with alerts and appropriate recommendations from experts.

These vendors generally offer a unified technology platform, advanced engineering and operations tools, and an integrated user interface to drive and sustain benefits across the enterprise.

At a growing number of process plants, remote monitoring keeps tabs on the advanced control applications already in place, enabling personnel to access data via the web to perform a detailed analysis of the process. Such remote monitoring can manage APC at multiple sites, resolving issues quickly and simply with a secure remote system connection.

Using proprietary technology, advanced control experts generate key performance indicators and diagnostics on how controllers are functioning. With this information, they create weekly reports that identify problem areas, such as the beginning of a breakdown in model relationships and degrading of application performance. This provides clear direction to onsite engineers on the changes required to optimize controller performance.

They work in the following way:

- *Evaluate current performance:* MPC vendor conducts reviews to assess health of systems, applications, performance and in-house skills.
- *Proactive support plan:* A sustainable remediation action list is created to address areas of vulnerability and to focus attention of in-house skilled resources.
- *Expert implementation:* MPC vendor performance specialists monitor outcomes, identify the root cause of issues, and recommend intervention by customer to sustain performance and long-term benefits.

References

Al-Ghazzawi, A., J. Anderson and T. Al-Soudani, (2003). "Avoiding the Failure: Achieving Sustained Benefits from Advanced Process Control Applications," proceedings of the Middle East Petrotech Conference, Manama, Bahrain.

Friedman, Y. Z. (1992). Avoid advanced control project mistakes. *Hydrocarbon Processing*, 71, 115–115.

Friedman, Y. Z. (1995). What's wrong with unit closed loop optimization? *Hydrocarbon Processing-Section 1*, 74(10), 107–116.

Friedman, Y. Z. (1998). More about closed loop optimization. A guest editorial, *Hydrocarbon Processing Journal*.

Friedman, Y. Z. (1999). Advanced control of ethylene plants: what works, what doesn't, and why. *Hydrocarbon Asia,* July August.

15

MPC Performance Monitoring

15.1 Why Performance Assessment of MPC Application Is Necessary

How to tell if your multivariable controller is doing a good job? Performance assessment is necessary to answer this critical question.

Multivariable controllers have been in use in manufacturing and production systems for many years. Multivariable controllers typically cost from $60,000 to more than $500,000, and they can deliver savings that are many times their cost. Yet the full benefits of these controllers are often not realized. Even worse, manufacturing sites may be completely unaware that the performance of the controllers is subpar. This section describes effective ways to measure and improve the performance of these powerful advanced controls.

Like any other investments in process industry, MPC applications also need payback calculations and performance assessment. Payback calculations are needed to convince management that putting money in MPC application is a profitable investment and can be considered as a top and attractive investment strategy.

Performance assessment is necessary to know the following:

1) Performance assessment after MPC implementation will give a true picture how much is achieved by a particular application as compared with initial study before implementation.
2) It will provide how much economic benefit that MPC can fetch.
3) From MPC service factor (i.e., percent of time MPC is online as compared to available time), it will be clear how much the developed MPC application gets operator confidence.
4) A periodical performance review will give an idea whether the performance of initial application is retained or gradually deteriorating.
5) A periodical performance review will also provide an idea for how much of the initial benefits are preserved over time and how much money is lost by not getting the full potential performance. This will help to justify periodical maintenance or overhaul of MPC application.
6) If performance is deteriorating, then from the trend of performance KPI it can be decided when to do maintenance or overhaul in MPC application.

Multivariable Predictive Control: Applications in Industry, First Edition. Sandip Kumar Lahiri.
© 2017 John Wiley & Sons Ltd. Published 2017 by John Wiley & Sons Ltd.

15.2 Types of Performance Assessment

There are two types of performance assessment:

1) Assess the MPC performance and calculate the benefits just after MPC implementation to justify the investment.
2) Periodically monitor MPC performance so that any deterioration in performance can be quickly detected and resolved.

Performance of model predictive control application (MPCA) can be evaluated by the following four categories.

15.2.1 Control Performance

- Whether MPCA is able to stabilize the plant operations
- Whether it is able to control all its key parameters within their desired range

15.2.2 Optimization Performance

- Whether MPCA brings down the standard deviation of its key economic parameters
- Whether it is able to run the plant at its limit or constraints to maximize economic benefit
- Whether it is able to hit multiple limits
- Whether it is able to bring down raw material or utility consumption or increase throughput

15.2.3 Economic Performance

- How much MPCA increases profit before and after implementation
- How much it increase production, decrease raw material and utility consumption, or decrease quality giveaway

15.2.4 Intangible Performance

- How much it saves operator time to monitor DCS. In some cases, most of the operator time is consumed to handle difficult control loops or untuned control loops.
- How much MPCA helps operator to handle those loops and relieve him
- How much operators love to keep MPCA online (can be known from on-stream factor or service factor)
- How much operator quality time shifted from monitoring level, temperature, pressure etc. to performance parameters like yield, efficiency, selectivity, constraints, and so on?

15.3 Benefit Measurement after MPC Implementation

It is necessary to calculate the tangible and intangible benefits after implementing MPC in the process plant. This helps to justify the MPC investment to top management. Multivariable control technology has two major benefits: reduced variability and operation closer to constraints. Reduced variability in the production process translates to energy and raw material savings, improved quality, and fewer production losses related to process trips. It is also the precursor for driving the production process closer to constraints to achieve greater overall production efficiency and profitability.

A full report of financial benefits, control benefits, and intangible benefits are usually carried out and presented to all relevant parties, including top management after MPC implementation. This may be the one-time activity, but can be repeated every year after implementation. Quantification of all benefits should be made wherever possible.

The report may consist of the following (but not limited to):

- Trends of key process variables before and after MPC implementation to show how MPC reduces the process variability and standard deviations.
- Trends of process parameters before and after to show how MPC helps to operate the process nearer to its constraints. This may be product quality parameters or safety related parameters.
- Trends of difficult to control process parameters before and after to show how MPC helps to improve control and stabilize the process.

Trends of followings may be furnished for pre- and postimplementation to show how MPC fetch economic benefits:

- Trends of utility consumption per ton of product (steam, water, etc.)
- Trends of raw material consumption per ton of product
- Plant throughput or capacity
- Catalyst selectivity or yield
- Key economic parameters
- Profit or operating cost

Report should end with clear-cut quantitative benefits in USD/year due to MPC implementations. It should also clearly indicate the payback period. It is a good practice to discuss the report with production and process engineers to have agreement regarding the benefit numbers before present it to top management.

The method to calculate the economic benefits should be the same as described in Chapter 3 during initial benefits calculations for consistency.

15.4 Parameters to Be Monitored for MPC Performance Evaluation

Periodic monitoring of MPC performance is needed to track its benefits. One problem that has hampered the progress of MPC throughout the industry is a lack of monitoring how the MPC schemes perform once they are commissioned.

The best way to tackle this issue is to develop quantifiable performance KPIs for MPC.

Assessing the performance of a multivariable controller requires producing metrics that reveal problems, impairment, and loss of benefits. An excellent review of performance assessment of MPC controller can be found in Jelali (2006).

Many researchers over the years (Al-Ghazzawi 2003; Jelali 2006; Paulonis 2003, Qin 1998, 2007 and 2012; Zhang 2007) have proposed different KPIs for multivariable MIMO controller performance assessment.

The following KPIs are recommended.

15.4.1 Service Factors

This is the percentage of time MPC is online versus total available time. Normally, it represents how much time operators keep the MPC switch on.

Monitoring service factors is done by historization of the scheme master tag. If MPC is kept switched on, the PV of that master tag is 100 and it will be zero if scheme is not working or switched off. There are two types of service factors that can be continuously tracked: one for the overall plant and the second for each controller or section of the plant. They can be summarized on a daily, weekly, or monthly basis.

A service factor more than 95 percent is desirable. If the service factor is reported below 90 percent, that means there are serious drawbacks to the scheme. Check the following if the service factor comes below 90 percent:

- Are MPC software-related issues making the MPC switch off every now and then?
- Is the operator switching off at a particular MPC application or controller?

Take it positively, as in most cases there is a valid reasons. The most common reasons are:

- Operators feel that MPCA cannot keep the process or a particular important parameters under control.
- MPCA drift the process parameter, which ultimately leads to safety or quality concerns.
- Operator feels that corrective or preventive actions MPC is taking are not correct. Operator counseling is needed in such cases.

Whatever may be the reasons, they must be addressed technically with professional manner. No MPC application will be be successful in the long run if plant operators do not support it.

15.4.2 KPI for Financial Criteria

The main purpose of MPC application is to increase profit margin of the plant. If this is not achieved, all other benefits soon become useless to top management. Thus, it is absolutely necessary to develop KPIs to measure financial benefits of implemented MPC scheme. A suitable profit margin criteria can be defines as follows:

$$Profit\ margin\ = \sum (Product\ flow_i X\ unit\ sales\ value_i)$$
$$-\sum raw\ material\ consumption_i\ X\ Unit\ value\ of\ raw\ mat_i$$
$$-\sum utility\ consumption_i\ X\ Unit\ value\ of\ utility_i$$

Utility should include all the utilities used in the plant which MPC can alter, such as team, power, instrument air, DM water, plant air, and chemicals used.

Good MPC scheme normally increase product flow, reduce raw material, and decrease utility consumption. Thus, it should increase profit margins as compare to no MPC scheme. This financial KPI should be generated and displayed as a separate tag in DCS so that everybody in the plant can see and monitor it.

Advantages of such tags are as follows:

- They give a clear idea of the value of MPC scheme, which can be shown to top management.
- In the long run, monitoring such KPIs provide an idea of whether the initial gain obtained after MPC implementation is diminishing with time or the initial gain is

retained. This will give an idea when a complete overhaul of MPC scheme including its functional design is necessary.

- It will also give an idea which utility or raw material consumption reduction gives maximum benefit. What are the contributions of each utility on overall profit margin?

15.4.3 KPI for Standard Deviation of Key Process Variable

This KPI represents the ability of the MPC scheme to maintain a process variable at its target.

One of the major purposes to implement MPC is to reduce variability of process parameters around given target, such as product quality variation at its specified target. Variability monitoring is accomplished by historization of the value of the following:

$$KPI_{variability} = \left|(SP - PV)\right|$$

Or it can be expressed as standard deviation as below:

$$KPI_{variability} = Standard\ devaition \left|(SP - PV)\right|$$

The value should be averaged for 7 days, once for when MPC scheme is working and once when it is not. When a scheme works well, it should be exhibit typical 50-75 percent reduction of standard deviation of key process variable. Process parameters to be monitored by this KPI are given in this section.

15.4.3.1 Safety Parameters

These parameters represent safety of the plant operation and should not be violated. For example, in ethylene oxide plant, reactor inlet oxygen concentration in cycle gas is kept at 8 percent (max) to avoid flammable mixture. This KPI will give an idea of how close MPC is able to keep the oxygen concentration to 8 percent.

15.4.3.2 Quality Giveaway Parameters

These parameters represent the quality of final product or quality of some intermediate stream that needs to be monitored closely. Example: Aldehyde content in ethylene oxide product should be maintained below 10 ppm to make the product on spec. Aldehyde content is measure by an online analyzer. This KPI will provide a measurement of how much variation is reduced by MPC on aldehyde content.

15.4.3.3 Economic Parameters

These parameters have profound effect on plant economics. Example: A multistage reciprocating oxygen compressor discharge flow is limited by the discharge temperature of the compressor, which depends on cooling water temperature of inter-stage cooler. In summer daytime, discharge temperature rises near to trip value and thus flow through the compressor cannot be increased further. In peak summer, flow has to reduce to avoid trip of compressor on high discharge temperature. But in nighttime and/or winter seasons when cooling water temperature is low, discharge temperature of compressor will come down. This opportunity should be exploited by well-designed MPC controller by increasing the oxygen flow through compressor. MPC should maintain the discharge temperature very near to trip value (say 1 °C below trip value) all the time and increase compressor load instead when opportunity exists. This KPI will give an idea of how close and how much time MPC is able to maintain the discharge temperature at its target value.

15.4.4 KPI for Constraint Activity

Another key objective of MPC is to operate the plant at its limit. As discussed earlier, this limit can be equipment limit (like high discharge temp of compressor), process limit (like high reaction temperature), quality limit (like maximum amount of particular impurity in product) or capacity limit (like percent flooding in distillation column). Constraints controllers are normally designed with soft constraint target such that when constraints are active, the desire is to keep the process variable at its limit, accepting small violations of that limit about 25 percent of the time. For that type of constraint controller, KPI should be percent of time constraints is active. Constraint activity can be monitored by historization of the following KPI.

$$KPI_{constraints} = \% \text{ of time } \left(\left\| (Limit\ value - PV) \right\| < 5\% \right)$$

Charts can be made to show the number, distribution, and identification of actively constrained parameters.

15.4.5 KPI for Constraint Violation

One of the key features of MPC is to operate the plant in such a way that all the constraints are obeyed—that is, any critical process parameters should not be operated outside its limit.

In addition to the soft target discussed earlier, there is also hard target, which plants do not wish to exceed. For those constraints, monitor the percent of time there is a violation.

$$KPI_{constraint\ violation} = \text{percent } of\ time \left(PV > Limit\ value \right)$$

It is advisable to monitor it before and after implementation. A well-tuned, well-designed scheme should be able to show a significant improvement in the frequency of violations.

15.4.6 KPI for Inferential Model Monitoring

Inferential model should be compared against synchronized laboratory value (or online analyzer) and deviations should be reported. Reliable inferential calculations should follow laboratory trend and the deviations should be less than 1 percent. If that is established for a 6 to 12 month period, then inferential calculations will quickly gain acceptance among plant operation people and be used more often.

$$KPI_{inferential\ model} = \left\| (Lab\ reading - Inferential\ model\ reading) \right\|$$

15.4.7 Model Quality

Good model is the building block of any MPCA. Verification of model quality can be done by measuring unbiased predictions and prediction errors. Low prediction errors for key variables will give an idea how well the models are built.

15.4.8   Limit Change Frequencies for CV/MVs

It is advisable to track number of times that operator limits are changed over a one-week period. Excessive changes indicate trouble.

15.4.9   Active MV Limit

MPC don't violate MV limit ever. This KPI will track the percentage of time MVs are hitting at their low or high limit. Suppose a particular MPC scheme objective is to reduce steam consumption in an evaporator train. Operator gives 90 and 110 MT/hr as steam flow low and high limits. MPC is expected to bring down the steam consumption over a period of time, and it should always hit the low limit of 90 MT/hr. At this point, operator should change the low limit to 80 MT/hr (say) to allow MPC to reduce steam further if scope exists. If this is not done, it will lead to potential loss of benefits.

$$KPI_{active\ MV\ limit} = \%\ of\ time\ \left(MV = high\ or\ low\ Limit\ value \right)$$

The purpose of this KPI is to warn or remind operators to change the limit to allow the MPC to play on a bigger ground. MPC will then try to reduce it further if the scope exists (i.e., by obeying all the CV limits).

15.4.10   Long-Term Performance Monitoring of MPC

All of the above KPIs should be historize and track over a one-year period to monitor MPC performance deterioration over a long horizon.

It is reported that initial investments of MPC delivers good benefit and performance. Then MPC benefits starts deteriorating without proper monitoring and maintenance. When operating in its best condition, a multivariable controller can return many times its original investment. Perhaps when the multivariable controller was first implemented there was an audit of the benefits achieved. The company made an effort to justify the original investment in the technology. There are many ways a multivariable controller can suffer performance losses. To name a few, performance of process equipment changes due to fouling, catalyst efficiency, wear and tear etc. Also addition of new equipment, change of pump impeller, cleaning of heat exchanger tubes, chocking of packing can significantly affect the step test model. These conditions will reduce the return on the original investment. This KPI should track this performance deterioration and will give an idea when to re step test or perform a model change activity. The best KPI for this is to monitor the economic profit margin as a percentage of initial profit margin gained just after implementation.

$$Profit\ margin\ as\ \%\ of\ original = \frac{Current\ profit\ margin \times 100}{Profit\ margin\ immediate\ after\ installation}$$

It should be trended over a year period and tracked periodically. Reactive and proactive maintenance of MPC scheme should be scheduled accordingly. Proper preventive and corrective maintenance should be planned to maintain the profit margin over a year, the same as achieved just after implementation.

To advocate effectively for funds to maintain the controller, point out the losses in benefits when the controller is not functioning optimally. We have discussed several technical measures to identify performance problems, but an economic indicator may well be the best tool to help justify when a company should conduct maintenance. One common way to establish the value added by the multivariable controller is to compare production profitability from a period when the multivariable controller was not in use to the current profitability while the controller is in use. Should that difference in profitability start to get smaller, investigate the reasons for that decline. When the reason for the difference is identified, use the difference in profitability to justify the expense of correcting the problem.

15.5 KPIs to Troubleshoot Poor Performance of Multivariable Controls

When the performance of MPC controller is below expectation, then it is necessary to develop some KPIs, which can be used to troubleshoot the problems quickly. Two levels of KPIs are needed. The first-level or higher-level KPI as described is to detect whether performance is deteriorating over time. Second-level or detail-level KPI is needed to troubleshoot the root cause once the performance deterioration is detected. These KPIs are not so obvious and should be formulated based on process. Main purpose of these KPIs is to assist troubleshooting if a performance deterioration is detected by the above KPIs.

Multivariable controller performance issues come from four general areas:

- Multivariable controller implementation
- Regulatory layer controls
- Operator actions
- Process changes and disturbances

Sometimes it is important to troubleshoot the problems if performance deterioration is detected. For troubleshooting the problems, some additional KPIs are needed, along with the top-level KPIs to pinpoint the root cause.

To ensure the performance of the full multivariable control system, we must have metrics or KPIs to detect and resolve each of the performance issues.

15.5.1 Supporting KPIs for Low Service Factor

As discussed earlier one early metric applied to multivariable controllers is "time in service" (on or off). This metric has proven to be of limited use, because functionality can be significantly impaired while the controller is still technically on. Measuring performance this way is akin to assessing an individual's performance based on how long the office light is on. Do not discard this metric completely, however; a low time in service is usually not good and should instigate an investigation to get to the root cause. But, if there are no supporting metrics to help set the direction of that investigation, it could be time consuming, involving operator interviews and trend analysis. Good supporting metrics help analysts get to the root cause sooner. Good supporting metrics are more important in the opposite scenario, where the multivariable controller has high time in service but other impairments exist.

Supporting metrics can be high-level metrics or detail metrics. A high-level metric is one that flags a performance issue. It can be a combination of several detail metrics. The previously discussed time-in-service metric can be made more useful by rolling the state of all the controller variables into the metric. If a low time-in-service condition occurs, it is easy to identify the controller variables that are responsible. A detail metric gives information on a specific behavior or condition. They are generally applied to individual controller variables; an example is "time at limit." An important requirement of a good metric is that it registers deviation from what is normal or optimal. Considering the time-at-limit metric, it would be good to know if being at a limit was good or bad, and this can vary from variable to variable. Good metrics alert us to a performance issue without making us scan historical trends and apply personal and possibly inconsistent interpretations to the information. When a performance issue is identified, a good metric provides information that leads us to the source of the problem.

15.5.2 KPIs to Troubleshoot Cycling

Multivariable controller model inaccuracy can also lead to cycling. As discussed before, the process itself can change over time. A detail metric like controller prediction error can help get to the root cause for this type of change. This is an indication that it is time to perform maintenance on the multivariable controller. Maintaining an average prediction error detail metric for each controlled variable permits the identification of the subset of variables most affected by the process changes. A trend of the average prediction error for a controlled variable provides some guidance on the direction a model update should take. For example, if the prediction error (Actual – Predicted) is consistently negative, this means one or more of the models affecting this controlled variable should decrease the gain. It may also be possible to address the over prediction with adjustments to model time constants. This may be less desirable, however, because it may necessitate additional adjustments to the overall controller time horizons.

The absolute value of the prediction error can help with two pieces of information. If the prediction error is alternating between over and under predictions, the average prediction error could end up near zero—leading us to assume there are no concerns. The average absolute value of the prediction error highlights the magnitude of the prediction error and provides an alert if the average prediction ends up close to zero. Taking the standard deviation of the prediction error is a measure of the dispersion of the error condition. It is not necessarily a bad situation when a controlled variable has a prediction error. A steady consistent prediction error causes little harm to controller performance. The opposite is true for a prediction error that is bouncing around. This condition could make it difficult for the controller to keep the variables within limits and very likely will reduce optimizing time. The prediction error standard deviation identifies this situation.

So important KPIs to troubleshoot cycling are as follows:

$$KPI_{prediction\ error} = \left|\left(Actual - Predicted\right)\right|$$

$$KPI_{prediction\ error\ std\ dev} = Standard\ deviation\left(Prediction\ error\right)$$

15.5.3 KPIs for Oscillation Detection

Another important metric is oscillation identification; after all, one of the purposes of a multivariable controller is to reduce variability. Oscillation condition is a high-level metric. In general, it needs some supporting metrics to qualify if a particular cycle is a problem. Using the amplitude of the cycle is one way to sort out small, insignificant behavior. Compare the amplitude of the cycling variable to the operating range the operator has allowed. If the amplitude is as large as the span of the operator limits, perhaps someone has overly constrained the controller, and it is just moving from lower bound to upper bound. Has the variance of that variable changed? Some multivariable controllers have an optimizing function that operates when there are free manipulated variables and no controlled variables are predicted to cross limits. This feature can drive the production process to a more economically attractive operating points. If a controlled-variables oscillation is reaching operator limits, the controller must leave the optimizing mode and return to enforcing the limits of the variables. This is inefficient and could be the cause of some cycling itself.

The period of the oscillation should also be considered. Long-period oscillation is less of a concern, and could be just due to the controller making adjustments for unmeasured disturbances. Another possible long-period oscillation is the multivariable controller responding to day-to-night temperature cycles. In this situation, that cycle could be economically favorable; perhaps the operation runs at a cooling or heating constraint most of the time. The cycle is observed as the multivariable controller takes advantage of the temperature change. It can reduce energy consumption or, more likely, increase production to take advantage of the atmospheric temperature changes. A common impairment to a multivariable controller that can cause a cycle is malfunctioning valves. A regulatory layer controller or valve problem can be seen as an oscillation period that is considerably smaller than the control horizon time. These cycles would also be observable at the individual loop level. Because a multivariable controller processes multiple inputs and outputs, the effects of a hardware-induced cycle can be magnified. It is a good practice to monitor regulatory layer control loop performance to quickly sort out the origin of a regulatory layer problem. It is possible that the problem is occurring on a regulatory layer controller that is not part of the multivariable control scheme.

$$KPI_{amplitude} = Amplitude \ of \ cycle$$
$$KPI_{period} = Period \ of \ oscillation$$

15.5.4 KPIs for Regulatory Control Issues

Regulatory layer control issues consist of controller tuning changes and measurement and valve problems. A multivariable controller relies on measurements collected by the regulatory layer. If those values are erroneous or upset (e.g., oscillation caused by valve sticking), they can cause less than optimal behavior from the multivariable controller. A multivariable controller is tuned to manage a production process. Regulatory layer controller tuning is embedded in the production process model the multivariable controller uses. KPIs to evaluate regulatory control were discussed in earlier chapters, and are not repeated here.

15.5.5 KPIs for Measuring Operator Actions

Another source of multivariable controller impairment comes from operator activity. This consists of actions taken to turn variables on or off or to change limits on variables. Some limit changes restrict the controller from achieving a more optimal operating condition; others can set up infeasible conditions, severely impairing the controller.

A good high-level metric is a count of times any variable is turned off or on, or a limit value is changed. Referencing this count against a norm alerts us to the possibility that the operators are having difficulty with the behavior of the multivariable controller. Drilling into the detailed metrics for the count values of the individual variables reveals what variable or part of the controller is of concern. Just because there was an excursion in operator activity does not mean the multivariable controller is impaired. Adding another piece of detailed information, like the available operating range of the variables, flags a situation where the operator change restricted the flexibility of the controller. Possibly the worst case is where a feasible operating point does not exist, and the multivariable controller simply saturates at several limits. A high-level metric to monitor the number of constrained variables and the individual time-at-limit metric for those variables would also flag an impaired controller.

A combination of the three metrics, number of changes, operating range, and constrained variables, provides a good screening tool to identify multivariable controllers that need attention. A person responsible for the performance of half-a-dozen or more multivariable controllers could spend quite some time looking through trends to determine if a controller is out of normal condition. It is much more efficient to have a screening tool point out specific multivariable controllers not meeting their expected performance metrics.

15.5.6 KPIs for Measuring Process Changes and Disturbances

The components of the production process (pumps, vessels, motors, and control systems) need to be evaluated and monitored to maintain operating performance. Bearings are lubricated, valves are repacked, and heat exchangers are cleaned to achieve their expected lifetimes and avoid sudden failures, accidents, and disruption to business. The hard or physical components of our systems generally get the care required; these are components that we can touch (bearing is too hot) or that we can see (leaks from a seal). There tends to be less attention to the soft components of the production process, because the metrics to assess condition are not as obvious. Multivariable controllers and control systems in general fall into the soft component category. It is similar to evaluating the condition of your home heating and cooling. If the thermostat setting (metric) is being achieved, is everything good? Not necessarily; the system might be turning on and off more frequently or for longer periods of time. The performance metrics need to be more sophisticated. Power consumption, outside temperature, and airflow together give a better picture of the condition of the system.

15.6 Exploitation of Constraints Handling and Maximization of MPC Benefit

MPC reduce process variability. But this alone cannot provide benefits. Set point has to change close to limit to get the real benefit. In most of the application, a well-designed MPC scheme is capable of reduction of 50 to 75 percent of standard deviation of key process parameters. MPC has an inbuilt model that acts both as feedback and feed-forward modes, it has future prediction capability, it can predict the future steady state constraint violation, and it takes small steps continuously to enable the process to run within their limit. It is the job of MPC model designers to evaluate which key parameters should be reduced and how. The idea is to reduce the variability of those process variables that can be translated to gain significant economic advantage or to increase process safety by reducing constraint violation or reduce quality giveaway. All these considerations should be tackled during the functional design and model building stage. Now postimplementation stage, it is the job of plant engineers and operators to change set point and exploit the benefit. This can be explained by Figure 1.5 in Chapter 1. Assume the trend in Figure 1.5 represent an impurity (say aldehyde content) in a product (instead of plant capacity in Figure 1.5). The maximum limit of aldehyde for on spec product is 100 ppm. Operators used to run the plant before MPC implementation at 60 ppm. The reason is simple: the variability was ±30 ppm. Operators want to control the aldehyde at 60 ppm (on an average) so that in extreme condition it will reach 90 ppm and does not make the product off spec. Operator tries to operate the plant in his comfort zone by keeping the safety margin. To keep the average aldehyde value at 60 ppm, operator may have use extra reflux flow and thus consuming more steam in distillation column. Now MPC engineer has to identify beforehand that there is opportunity for improvement and design the controller in such a way that this variability can be reduced. After implementation, process variability is reduced to ±10 ppm, as shown in the figure. Now after observing this reduction of variability of quality parameter, it is the job of plant engineers to enhance the level of limit from 60 to 80 ppm. Benefit is twofold: It will allow to reduce reflux and steam in product distillation column and also it will give slightly more product flow as impurity in the product will increase.

So, to exploit the benefit of MPC postimplementation stage, the following actions are recommended:

- Monitor the trend of all key economic and safety process parameters before and after implementation.
- Find out quantitatively how much reduction of standard deviation of key parameters are achieved.
- Explore how much limit change is possible due to reduction of process variability.
- Determine how much operator safety margin can be cut realistically to extract economic benefit from process.
- Change the limit and keep monitoring.
- Calculate the associated economic benefit for shifting of limit says steam reduction due to reflux reduction in the above example.

References

Al-Ghazzawi, A., J. Anderson and T. Al-Soudani, (2003). "Avoiding the Failure: Achieving Sustained Benefits from Advanced Process Control Applications," proceedings of the Middle East Petrotech Conference, Manama, Bahrain.

Jelali, M. (2006). An overview of control performance assessment technology and industrial applications. *Control Engineering Practice*, 14(5), 441–466.

Paulonis, M. A., & Cox, J. W. (2003). A practical approach for large-scale controller performance assessment, diagnosis, and improvement. *Journal of Process Control*, 13(2), 155–168.

Qin, S. J. (1998). Control performance monitoring—a review and assessment. *Computers & Chemical Engineering*, 23(2), 173–186.

Qin, S. J. (2012). Survey on data-driven industrial process monitoring and diagnosis. *Annual Reviews in Control*, 36(2), 220–234.

Qin, S. J., & Yu, J. (2007). Recent developments in multivariable controller performance monitoring. *Journal of Process Control*, 17(3), 221–227.

Zhang, Q., & Shaoyuan, L. (2006). Performance monitoring and diagnosis of multivariable model predictive control using statistical analysis. *Chinese Journal of Chemical Engineering*, 14(2), 207–215.

16

Commercial MPC Vendors and Applications

16.1 Basic Modules and Components of Commercial MPC Software

After many mergers and acquisitions in the late 1990s, there are very few MPC vendors available today who can be considered as real technology providers and licensors of MPC technology. Clearly, DMCplus of AspenTech, RMPCT of Honeywell, and SMOC of Shell Global solutions emerged as winners of this stiff competition in MPC technology and holds the majority of MPC business share globally. Figure 16.1 provides some major companies and their products available today in linear MPC technology.

Major MPC applications present on the market during the 2000–2015 were:

- AspenTech
- Honeywell
- APCS (Adaptive Predictive Control System): SCAP Europa
- DeltaV MPC: Emerson Process Management
- SMOC (Shell Multivariable Optimizing Controller): Shell
- Connoisseur (Control and Identification package): Invensys
- MVC (Multivariate Control): Continental Controls Inc.
- NOVA-NLC (NOVA Nonlinear Controller): DOT Products
- Process Perfecter: Pavilion Technologies

To support the market needs, some of these technologies started to include nonlinear MPC, while most of them were still linear MPC. An excellent survey of industrial model predictive control is given in Qin 2003; Qin 1997. History and development of model predictive control, can be found in Ruchika 2013.

MPC is a matured but emerging technology. Though last five years, there is no breakthrough development in core MPC technology but these MPC vendors come up with more software packages, which helps to implement and monitor MPC technology in shop floor. These MPC vendors are now offered a full range of software package, as described below.

16.1.1 Basic MPC Package

- System identification module for offline model building
- PC-based offline simulation package
- Online control module

Multivariable Predictive Control: Applications in Industry, First Edition. Sandip Kumar Lahiri.

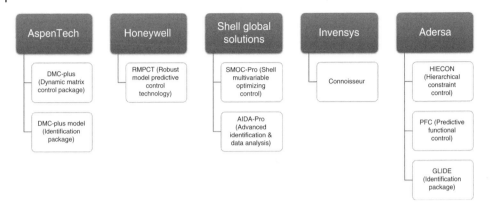

Figure 16.1 Major linear MPC companies and their products

- Soft sensor module (Also called quality estimator module)
- Control performance monitoring and diagnostics software

The core technology and the basic features of these software of various commercial MPC vendors are not very different. All of them have special two or three unique technical features, but the basic features are very common among them. They look cosmetically different.

Figure 16.2 shows the basic structure of the MPC software. At the heart of MPC project is the process plants with all of its PID controllers.

Plant is connected with DCS/PLC systems, which scan the plant every seconds. MPC module starts at the supervisory level above DCS. MPC online software package normally interacts with DCS via interface. The following sections describe various component of MPC software and their common features available in commercial software. It should be noted that various features discussed here are usually available in all commercial software in various looks and forms. However, to know exact features of software by different MPC vendor, readers are requested to refer their respective website.

16.1.2 Data Collection Module

It starts with data collection module. Purpose of this module to collect real-time data directly from DCS or historical data from data historian. Usually these data collector can collect data either at the DCS level (usually compliant with Honeywell AxM/TPS, Foxboro IA, and Yokogawa CS) or from a process computer system with SETCON, Info Plus, PROSS 2 or OSI's PI. This data collection module used extensively during step test and soft sensor building.

16.1.3 MPC Online Controller

This is the brain of the whole MPC package. The online controller incorporates all functions required for the operation of the controller in real time. This includes signal validation, initialization, mode shedding, standard operator and engineer interface, and controller engine. Usually, the same online controller is available in versions running on different platforms, either at the DCS level (Honeywell AxM/TPS, Foxboro IA, and Yokogawa CS) or in a process computer system with SETCON, Info Plus, PROSS 2 or OSI's PI.

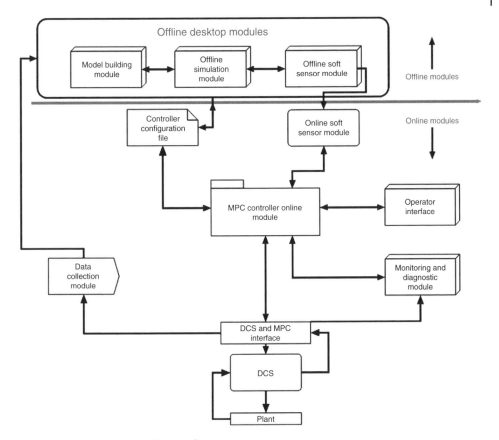

Figure 16.2 Basic structure of MPC software

16.1.4 Operator/Engineer Station

Normally, operators and engineers interact with the online controller through this interface. Operators enter set points, high–low limits of CVs and MVs; engineer enters different tuning parameters through this station. They can see, start, stop the MPC application, and drop a CV or MV through this platform. MPC software can install these custom base stations in dedicated PCs inside the control room at the side of DCS.

16.1.5 System Identification Module

The heart of any MPC package is the system identification module, which is also known as a model-building module. There are many modeling techniques available in the modules of different vendors. This module is considered as an earmark of the MPC technology advancement, and over the years sophisticated modeling technology has been added in this module. A comparison of MPC identification technology by different vendors is shown in Figure 16.3. MPC identification technology is evolving every year and readers should refer to the respective websites of MPC vendors for their latest offerings in identification technology.

Common features of system identification or modeling module are given in the following sections.

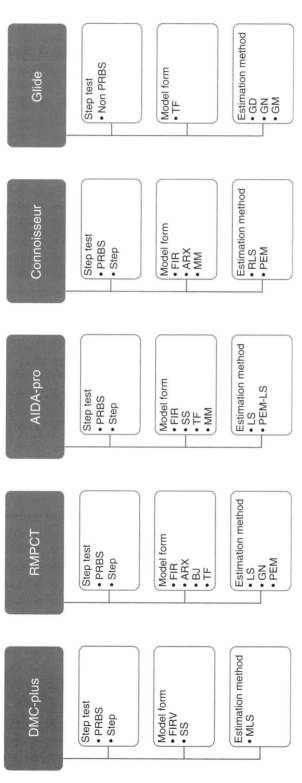

Figure 16.3 Comparison of different MPC identification technology*

*Where different terms are given below:

Model form: Finite impulse response (FIR), velocity FIR (FIRV), Laplace transfer function (TF), state-space (SS), auto-regressive with exogenous input (ARX), Box–Jenkins (BJ), multimodel (MM).

Est. method: Least-squares (LS), modified LS (MLS), recursive LS (RLS), subspace ID (SMI), Gauss–Newton (GN), prediction error method (PEM), gradient descent (GD), and global method (GM).

16.1.5.1 Different Modeling Options

In most of the system identification module models can be obtained in various ways. Two very common ways are: Finite impulse response (FIR) fit followed by a parametric reduction and direct parametric fit of various models to data.

16.1.5.2 Reporting and Documentation Function

- *Modeling:* Module is designed to expedite the tasks performed during the course of a model identification analysis. Modeling module of almost all the major vendors has a comprehensive reporting, project documentation, and bookkeeping functions.
- *Data import:* Modeling module includes a versatile data import capability, designed to work with a variety of formats. Plant data can also be retrieved directly from data historians.

16.1.5.3 Data Analysis and Pre-Processing

- *Modeling:* The module has powerful trending capabilities. Data segmentation and bad data handling can be performed graphically. A wide variety of built-in calculations are also included.
- *User-friendliness:* A modeling module workspace is a collection of process data, identification studies, and assembled models. It serves as a single-project archive and simplifies controller model maintenance. The modeling module is normally integrated with vendor other process control software suite, sharing of objects between the various products, and also in its underlying modeling philosophy.
- *Statistical analysis capability:* Modeling module normally provides statistical 95 percent confidence intervals for the estimated step responses and model parameters. Residual trends and goodness-of-fit tests, such as, residual auto- and cross-correlation analysis, can be used to check the quality of models.
- *Model validation and selection:* Modeling module can be used to build several different models and their predictions can be compared against actual data. Models are easily constructed by mixing and matching various identification results. Final models are exported to online MPC control module for real time configuration.
- *Incorporating process knowledge:* It is not common feature, but some of the vendor's package has this capability. In some of vendor package users can perform constrained model identification using prior process knowledge. This can include setting gain, time constants, and dead time within bounds.
- *Grey-box and black-box modeling:* The knowledge of cause–effect relationships can be incorporated during the process of model identification (grey-box modeling). There are also special features in some of the packages.
- *Unmeasured disturbance modeling:* Not all the vendor packages has this capability. SMOC has this feature in detail. Rejecting unmeasured disturbances is one of the main functions of closed loop control. Unmeasured disturbances include measurement noise, sensor drift, changing ambient identification, and some controller parameters (e.g., gains), are incorrectly estimated. This could result in overly aggressive control action. Modeling module allows factoring in unmeasured disturbance effects during model identification.

16.1.6 PC-Based Offline Simulation Package

Offline simulation module is a Windows-based package, integrated with system identification module, to design, test, and build MPC controllers. The process dynamic model is generated by identification modules. Alternatively, the model may be built using the Graphical Model Builder. PC-based offline simulation module allows the user to simulate the controller behavior to test the tuning, robustness to model errors and optimization performance. The final stage in the offline design is the creation of the MPC controller file, for use by the online controller.

The dynamic models used by a MPC controller can be built either in a matrix format using a Microsoft Windows based package and/or a flow sheet using the Graphical Model Builder available in the most of the offline package. Offline configuration tools are also provided to design, test, and build MPC controllers. The offline package generates the online controller implementation file. Offline package features a powerful simulation environment to test the controller behavior. Scenario based simulations can be used to test tuning, robustness to model errors and optimization performance.

16.1.7 Control Performance Monitoring and Diagnostics Software

Usually these software suites are used for monitoring all base level and commercially available multivariable control systems. They also feature offline applications that can be used to perform advanced diagnostics on all single output and multivariable control loops, and another for analyzing PID loops only. Nowadays, this technology is being used extensively in many refineries and chemical sites around the world.

16.1.7.1 Control Performance Monitoring

It is generally recognized that model predictive control (MPC) generates large economic benefits. The achieved benefits are a direct function of the control system utilization and of its actual performance. Base layer controls, instruments, and analyzers must also perform correctly to ensure high APC utilization and profit. Control performance deteriorates over time due to changes in feed qualities, plant dynamics, new control objectives, new operating modes, and many other factors.

The advanced monitoring and diagnostics tool constantly monitors the control applications performance, model errors and other metrics. Dynamic tuning of an online controller can be carried out as well. There is also an interactive performance analysis tool that can help engineers to identify the bad actors. Once the bad actors are found, they are fixed to sustain good control performance, securing the MPC-generated benefits.

16.1.7.2 Basic Features of Performance Monitoring and Diagnostics Software
- Usually, any type of commercially available multivariable controller can be monitored.
- Multiple input data formats and historian systems are supported, for example, by PI, InfoPlusX, ExaQuantum, PHD, and IP21.
- Fast calculation and data handling allows a large number of loops to be analyzed in a short period of time.
- Diagnostic measures and statistics, such as closed-loop speed of response, can be compared with user-defined benchmarks.

- The technology and the user interface are easy to use and understand, and comprehensive online help is included.
- Loop reports can be shared using standard MS Windows applications.
- Automatic, exception-based performance reports distribution via email.
- PID tuning is available based on the desired response time.

Normally this software package consists of three main application

- Offline module for diagnosis
- Online for performance monitoring
- Web server module for viewing and reporting

16.1.7.3   Performance and Benefits Metrics

Large numbers of performance metrics are calculated online periodically. Some of the very common metrics are given below:

- *Lost opportunity:* Lost opportunity is the difference between the maximum daily benefit and the realized daily benefit associated with process or controller downtime.
- *% uptime:* The amount of time that a particular controlled variable (CV) is "on control" during the day.
- *% in-service:* The amount of time during the day that a CV was operational or available for control.
- *% in-compliance:* The amount of time during the day that a CV was on control and within a prespecified tolerance of its set point or set range.

16.1.7.4   Offline Module

The offline package can be used to analyze the performance of single-input, single-output controllers and multivariable controllers. The offline package uses statistical methods to extract the essential control performance information from a loop's set point, process value, and output trends. This saves the control engineer from the inefficient and time-consuming process of viewing hundreds of raw data trends in detail.

16.1.7.5   Online Package

The online module is a client/server system for monitoring plant-wide controller performance. It has been designed to operate on a typical business network.

16.1.7.6   Online Reports

1) *Overview summary report:* A summary of control performance for the entire facility, with uptime, compliance, and in-service statistics presented for each process unit.
2) *Bad actors:* Two lists of the worst-performing CVs in the facility ranked by consecutive days of not meeting objectives. The first list applies to CVs that are in-service, and the second list applies to CVs that are out-of-service.
3) *Poorly controlled control variables:* A list of the controlled variables in a unit that are not meeting user defined control performance objectives.
4) *On-demand reports:* Customizable summary reports consisting of all control performance and benefits statistics over a user-defined time period.

16.1.8 Soft Sensor Module (Also Called Quality Estimator Module)

Soft sensor module is for design and implementation of inferred measurements for process monitoring and control.

A soft sensor module inferred measurement is typically calculated using selected process variable such as temperature, pressure, and flow.

The soft sensor module technology is used extensively in many refineries and chemical sites around the world. Typical applications in the hydrocarbon industry include: ASTM boiling point temperatures, cold properties, flash point, Reid vapor pressure (RVP), Road Octane Number (RON), Impurity content in various streams, Melt index for polymerization processes.

Soft sensor module is used essentially in two different configurations:

1) As a standby of online analyzer: The inferred measurement provides a fast-responding continuous value of the property of interest, whereas the analyzer measurement often has a large time delay and/or is discontinuous. In addition, the inferred measurement provides a backup value when the analyzer fails or is being calibrated.
2) With laboratory update mechanism: the inferred measurement value is periodically updated on the basis of results from the laboratory. The inferred measurement can be seen as a replacement to an online analyzer, when it is not justified economically or when it is not technically feasible.

16.1.8.1 Soft Sensor Offline Package

The offline package is a fully Microsoft Windows-based graphical package for fast track design and maintenance of inferred measurements.

It includes the following functions:

- *Data analysis and pre-processing:* This function allows the user to import, display and perform statistical analysis of data sets of process variables to be used for modeling purposes.
- *Inferred measurement modeling:* The modeling phase consists in selecting the calculation inputs, selecting the model type, estimating the model parameters and validating the resulting model. Various options are available.
- *Usually linear models (including PLS, PCA) and neural network types:* The user does not need to have detailed technical knowledge on modeling. The modeling, including inputs and delay selection, can be done fully automated in a few, but reliable, steps.
- *Inferred measurement update design:* This part of the package allows the user to test and tune several different options available for updating the prediction model from online analyzer or laboratory measurements. The available options include SPC rules (Statistical Process Control), Kalman filtering and a simple bias update mechanism.

In some of the vendor package special type of neural network models (RBF networks) has been chosen for use in this module. RBF networks, unlike standard neural networks, allow the application of a method to automatically select the network structure, and can be combined with the Kalman filtering prediction model update option.

Usually these modules can interact with major DCS platforms (Foxboro, Honeywell, Yokogawa), and also on a number of real-time control process computer systems (Pross2, Infoplus and Setcon). Compared to conventional bias update mechanisms, Kalman filtering (especially available in Shell RQEPro) provides a smoother prediction, has proven to give significant additional accuracy, improves the reliability of the online inferred measurement, and reduces maintenance requirements. The offline package generates a model file, including all parameters required for the online implementation of the inferred measurement.

16.1.8.2 Soft Sensor Online Package

Online package includes the full range of flexible and configurable tools required for the robust and reliable online implementation of inferred measurements.

The main features include the following:

- Extensive online analyzer signal checking, including out-of-range handling, frozen value detection and spike rejection
- Operator lab interface with entry consequence preview
- Inputs processing (dynamic compensation, nonlinear transformation, combination of variables, filtering)
- Prediction calculation and output processing (clamping, asymmetrical filtering)
- Full model updating using various options like Kalman filtering (either on basis of online analyzer or on basis of laboratory results), input checking of reference laboratory result prior to model prediction update
- Various techniques for outlier detection
- Optional steady state detection
- Robust handling of uncertain and varying delay of the quality reference measurement,
- Operator graphical interface and engineer graphical interface
- Messaging and alarming

16.1.8.3 Soft Sensor Module Simulation Tool

Simulation tool is an intuitive, user-friendly offline simulation package for testing the performance of Soft sensor module applications. It provides process control engineers the ability to tune soft sensor module applications created, using the soft sensor module offline package.

This simulation provides troubleshooting tools for existing soft sensor module applications. When used as a maintenance tool, this module can monitor model update frequency, prediction errors, quality measurement indicator (QMI) spike filtering, and other tuning parameters in the application.

16.2 Major Commercial MPC Software

This section describes development history and features of three major MPC player namely Aspen DMCplus, Shell Global solutions SMOC, and Honeywell's RMPCT. Note that all the features describe below were collected from open source literature and may not be fully describe the current features and offerings by the respective MPC vendor. This section divided into following subjects:

- Brief history of development of each MPC technology
- Briefly describes the product offerings of each vendor with some of their uncommon features (common features are already described above)
- Distinctive feature of their respective technology with current advancement

16.3 AspenTech and DMCplus

AspenTech is considered as a major MPC vendor with a large chunk of market share across the globe. They represents pioneer MPC technology provider and have expertise in implementing MPC in almost all the major refinery across world.

16.3.1 Brief History of Development

DMC technology first appeared in 1979 and represented the first generation of MPC technology. DMC technology is one of the pioneers of MPC technology developed by Shell Oil in the early 1970s, with an initial application in 1973. Cutler and Ramaker are considered the fathers of dynamic matric control (DMC) technology after their presentation of an unconstrained multivariable control algorithm at the 1979 National AIChE meeting (Cutler & Ramaker, 1979) and at the 1980 Joint Automatic Control Conference (Cutler & Ramaker, 1980).

Key features of the initial DMC control algorithm include:

- Linear step response model for the plant
- Future plant output behavior specified by trying to follow the set point as closely as possible
- Optimal inputs computed as the solution to a least squares problem

The linear step response model used by the DMC algorithm relates changes in a process output to a weighted sum of past input changes, referred to as input moves. For the SISO case, the step response model looks like equation 16.1:

$$y_{k+j} = \sum_{i=1}^{N-1} S_i \Delta u_{k+j-i} + S_N \Delta u_{k+j-N} \qquad [16.1]$$

By using the step response model one can write predicted future output changes as a linear combination of future input moves. The matrix that ties the two together is the *dynamic matrix*.

The dynamic objective of a DMC controller is to drive the output as close to the set point as possible in a least-squares sense with a penalty term on the MV moves. This results in smaller computed input moves and a less aggressive output response.

16.3.1.1 Enhancement of DMC Technology to QDMC Technology in 1983, Regarded as Second-Generation of MPC Technology (1980–1985)

Original DMC algorithm is enhanced with quadratic program (QP) to handle input and output constraints. QDMC algorithm first appeared in a 1983 AIChE conference paper (Cutler, Morshedi, & Haydel, 1983). Garcia and Morshedi (1986) published a more comprehensive description several years later.

The QDMC algorithm can be regarded as representing a second generation of MPC technology, comprised of algorithms, which provide a systematic way to implement input and output constraints.

Key features of the QDMC algorithm include:

- Quadratic performance objective over a finite prediction horizon
- Optimal inputs computed as the solution to a quadratic program

16.3.1.2 Introduction of AspenTech and Evolvement of Third-Generation MPC Technology (1985–1990)

DMC group was separated from Shell Oil and developed improved DMC algorithm by the name of QDMC and was later bought by AspenTech,

During the 1980s, MPC technology slowly started to prove the results to gain wider acceptance during the 1990s. During the 1980s, AspenTech concentrate to improve the original algorithm to tackle real-life shop floor control problems of refinery. Engineering teams continued the development of MPC algorithms and brought new implementations for improved handling. AspenTech was focusing how to improve handling of the constraints, fault tolerance, objective functions and degrees of freedom in their algorithms.

Milestone: In 1986, AspenTech released first commercially available model predictive control software. Today, Aspen DMCplus is considered as one of the best advance control product.

16.3.1.3 Appearance of DMCplus Product with Fourth-Generation MPC Technology (1990–2000)

The late 1990s were characteristic of MPC reaching its peak. In parallel with the market growth, the competition between the vendors was also reaching its peak during late 1990s.This was the time when major merging and acquisitions of companies started with the aim to control the market. AspenTech and Honeywell got out as the winners of this phase.

In the era of 1990–2000, increased competition and the mergers of several MPC vendors have led to significant changes in the industrial MPC landscape. In early 1996, Aspen Technology Inc. purchased both Setpoint, Inc. and DMC Corporation. This was followed by acquisition of Treiber Controls in 1998. The SMCA and DMC technologies were subsequently merged to create Aspen Technology's current DMC-plus product.

16.3.1.4 Improvement of DMCplus Technology for Quicker Implementation in Shop Floor, Regarded as Fifth-Generation MPC (2000–2015)

The consolidation of vendors continued over the last few years as well. AspenTech and Honeywell however still managed to preserve their leading role on the APC market. Today, we are witnessing a further technology development which is not so much focused on improving the algorithms, but to improve the development steps. The focus is put to make those steps smoother, faster and easier, both for the developer and for the client and do as much as possible remotely.

Aspen Technology has introduced a set of innovative technologies that are having a significant impact on APC methodology, applications development, benefits and sustainability. Closed loop auto step testing, seed model generation, adaptive modeling approach are some of the new innovation done to reduce the MPC implementation time.

Milestone: Aspen ONE software is released in 2004, which brings together integrated suite of optimization products across engineering, manufacturing, and supply chain function.

In 2012, AspenTech introduces Adaptive Process Control technology. Now part of Aspen DMC3, it enables process manufacturers to sustain and maximize benefits of advance process control by making maintenance a continuous process.

16.3.2 DMCplus Product Package

Now DMCplus comes with a comprehensive product package where different products are available under same umbrella, namely Aspen ONE (Figure 16.4). Good overview of DMC controller and its various features can be found in DMC Corp 1994; Lundström 1995; Muske 1993). As MPC technology is constantly evolving over the years, readers are requested to refer the respective website of MPC vendor to get the latest information and technology offerings.

16.3.2.1 Aspen DMCplus Desktop

Aspen DMCplus Desktop is composed of an integrated suite of three programs:

- DMCplus Model offers multiple model identification algorithms plus model prediction, model uncertainty, and cross-correlation features for model analysis.
- DMCplus Build makes controller configuration and maintenance easier with detailed, context-sensitive help and an automatic configuration validation wizard.
- DMCplus Simulate enables interactive evaluation and testing of controller performance in the face of model mismatch and process measurement noise.

16.3.2.2 Aspen DMCplus Online

Aspen DMCplus Online consists of two standard layers and an optional third layer:

- DMCplus Control is the online controller program and performs input validation, steady-state target calculation, and dynamic move calculations.
- DMCplus Connect links the DMCplus controller with the process unit.
- DMCplus Composite is an optional program to allow a mix of slow/fast dynamics and larger-scale solutions.

Key Features
- Integrated data collection, viewing, conditioning, slicing, and transforms
- Integrated model viewing, merging, convolution, and transforms

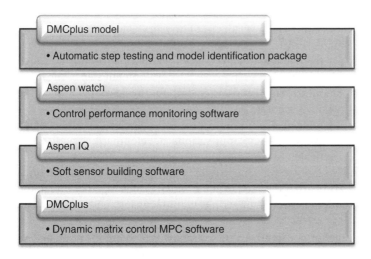

DMCplus model
- Automatic step testing and model identification package

Aspen watch
- Control performance monitoring software

Aspen IQ
- Soft sensor building software

DMCplus
- Dynamic matrix control MPC software

Figure 16.4 DMCplus product package

- Integrated sub-space model identification
- Integrated controller configuration and tuning
- Integrated simulation running against either a model or historical data
- Integrated tag browser
- Tag templates to simplify configuring plant IO connections
- Real-time online server

16.3.2.3 DMCplus Models and Identification Package
Aspen ONE Advanced Process Control provides three different controller formulations in a single package.

1) The industry-leading Aspen DMCplus® formulation based on the FIR model form (finite impulse response)
2) Linear MIMO State Space (multiple input, multiple output)
3) Nonlinear MISO State Space (multiple input, single output)

Switching between formulations is accomplished with a simple mouse click.

The modeling tools are implemented in a modern (.NET) drag-and-drop environment for model building, simulation, configuration, and deployment. The entire control application can be built, simulated, and deployed within Aspen ONE Advanced Process Control, providing a true competitive advantage with a seamless interchange between the three industry-leading control formulations.

16.3.2.4 Aspen IQ (Soft Sensor Software)
Inferential model types include FIR, PLS, fuzzy PLS, hybrid neural net, monotonic neural net, linearized rigorous model-based, and custom equations.

Key Capabilities
- Built-in steady-state detector
- Can be used stand-alone or integrated with Aspen DMCplus and Aspen InfoPlus.21®
- Provides both analyzer and lab model updating
- Wide range of DCS and information system interfaces available
- Enabled for remote monitoring
- No code generation or programming required

16.3.2.5 Aspen Watch: AspenTech MPC Monitoring and Diagnostic Software
Maintaining APC applications at peak performance requires three activities working in concert:

- *Detection:* Aspen ONE Advanced Process Control provides a full set of features for monitoring control applications, PID loops, and production assets. The three controller configurations are supported with preconfigured sets of KPIs that provide out-of-the-box controller performance monitoring. These KPIs are automatically preconfigured when you create a controller in the Aspen Control Platform. Users can add KPI definitions to provide additional insight into process performance and product quality. KPIs are supported with built-in calculation scripting features, greatly simplifying the task of creating and managing operational KPIs.
- *Diagnostics:* The most common cause of APC performance degradation occurs when the controller model no longer matches the actual plant performance. This mismatch

occurs as plant equipment wears or as process improvements are made. The standard KPIs for models help to quickly determine the source of model mismatch—including the ability to identify specific MV/CV pairs within the model that are at the root of performance degradation.

- *Corrective action:* Achieving maximum benefits from APC applications requires high service factors. When controllers do require revamps, AspenTech's sustainability features help to quickly collect new process data, clean that data in preparation for building new models, and provide automation to quickly iterate through candidate models.

16.3.3 Distinctive Features of DMCplus Software Package

16.3.3.1 Automating Best Practices in Process Unit Step Testing

AspenTech's automated step testing features simplify and streamline the task of collecting process data for model construction. Delivered model data have higher information content than manual testing while reducing the overall engineering efforts.

Key Capabilities

- Controller to enforce MV limits while keeping CVs within bounds. Generates a constrained step test (e.g., valve limits, column qualities are honored) Larger steps are possible.
- Operating points for step testing can be defined and modified by operations
- Allows any MV to be paired with another MV. Both are stepped in a given ratio in order to obtain more informative test data. Data allow identification algorithms to determine gain ratios and RGA numbers of nearly colinear submatrices
- Specify preferred first step direction for test moves.

16.3.3.2 Adaptive Modeling

Aspen Adaptive Modeling automates the maintenance lifecycle of a controller by providing the ability to collect historical data, automate calculations for data cleaning, schedule online model quality assessments, and assess model quality, model diagnostics, and online model identification for standard and custom KPIs.

Key Capabilities

- Integration with Aspen Watch Performance Monitor for historical data collection
- Ability to customize calculations for automated data cleaning
- On-demand or automated model quality assessments—assessment of model gain accuracy
- Library of model assessment KPIs
- Carpet plots for viewing model performance KPIs
- Model diagnostic tools
- On-demand or automated online model identification
- Subspace model identification
- Online comparisons of model ID cases
- Easy export of new models into updated controller

16.3.3.3 New Innovation

Aspen Technology has introduced a set of innovative technologies that are having a significant impact on APC methodology, applications development, benefits, and sustainability. A traditional Aspen DMCplus® application pushes for full LP optimization that exploits all available degrees of flexibility. However, when a large mismatch exists between the plant and the controller model, the LP may exhibit what is called *flipping* behavior.

This phenomenon drove elongated project cycles as heavy plant testing was required in order to develop highly accurate models before the application could be deployed. This often resulted in many months elapsing before the application could be put online to start generating operational benefits.

The new technology, Adaptive Process Control, is designed to solve these problems (Figure 16.5). Adaptive optimal control is discussed in detail by Bitmead (1990).

First, Adaptive Process Control maintains the LP steady-state target in an "optimum area" instead of an optimum point. Hence, optimization and stable behavior are achieved even in the presence of a very large model mismatch.

Second, Adaptive Process Control can utilize what are called *seed models* that are built from pre-test and/or purely historical process data or from an existing rigorous plant model. This means that—coupled with the robust optimization behavior in the presence of a model mismatch—the controller can be deployed with these less refined seed models, and as a result, in a much shorter period of time, begin accruing benefits while the model is improved online by the adaptive technology.

Third, once deployed, the acquisition of process data to refine the seed model is completed using small perturbation background step testing while the controller is online. The underlying enhancements to the model identification algorithm enable the effective use of significant amounts of closed-loop data that can be collected

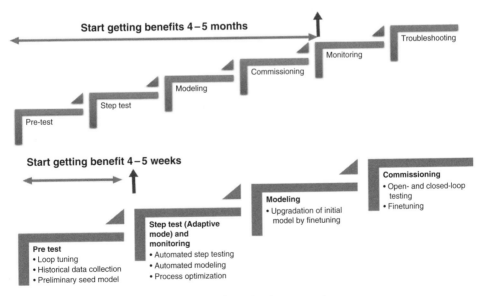

Figure 16.5 MPC project outline: Conventional vs. adaptive approach

while a user-specified level of emphasis on optimization is carried out by the system. That specification enables the user with the ability to make trade-offs between the speed of developing updated models and the degree of benefits lost.

The collective set of innovations within Adaptive Process Control is driving a radical change in the economics of APC applications. While it has been shown that the technology solves many of the maintenance issues with traditional APC solutions, it is also having a dramatic impact on the development of new APC applications.

Aspen shows how Eni's Livorno Refinery was able to use APC to start accruing benefits from two hydro treater units after just six weeks from the start of the project.

16.3.3.4 Background Step Testing

When in adaptive control mode, the optimization algorithm recalculates the steady-state (LP) objective function at every time step and stores this cost function (say, Cost1). It then calculates the cost function at the current MV targets (Cost2) with min move setting (constraint control), compares the two cost function values, and calculates $\Delta Cost = Cost1 - Cost2$. If $\Delta Cost$ is larger than the user-specified tolerance, the Aspen DMCplus engine will generate a move plan and push the process toward the new optimal solution.

In Figure 16.6, a process is shown that is bounded by constraints on six CVs. At a particular time, say an optimum point is situated at the intersection of two active CV constraints, as shown in Figure 16.6. The technology is trying to keep the current operating point inside the blue triangle instead of attempting to reach the optimum point at the corner. When inside the triangle, the engine will generate step moves for the requested manipulated variables. This essentially means when the current operating point is very near to optimum point, instead of trying to push it further, it will utilize the opportunity to do a step test. This mode of operation, where the unit is tested and optimized, is called *adaptive control mode.*

For application normally run in adaptive control mode in first few months, the unit has been optimized and tested in a nondisruptive way, implementing small steps on

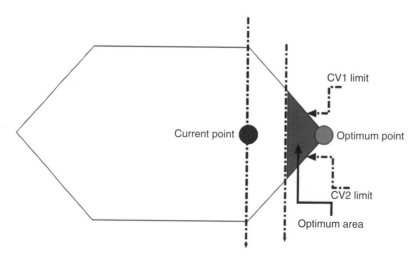

Figure 16.6 Optimization in adaptive control mode

each manipulated variable. Given the small steps size (even 10 times smaller than usual step test moves), operators hardly noticed that a step test was ongoing. They simply saw the benefit of a MPC controller in action.

16.4 RMPCT by Honeywell

16.4.1 Brief History of Development

RMPC algorithm by Honeywell came into market as third-generation MPC software in 1990s. From 1990 to 2000, increased competition and the mergers of several MPC vendors led to significant changes in the industrial MPC landscape. In late 1995, Honeywell purchased Profimatics, Inc. and formed Honeywell Hi-Spec Solutions. The RMPC algorithm offered by Honeywell was merged with the Profimatics PCT controller to create its current offering, called RMPCT. After that, Honeywell, with its RMPCT software, clearly emerged as an industry leader, with a major share in world MPC business (Figure 16.7).

16.4.2 Honeywell MPC Product Package and Its Special Features

A good overview of RMPCT Profit Controller and its various features can be found in Bowen (2008); Honeywell Inc. (1995); Lu (1997); and MacArthur (1996).

16.4.3 Key Features and Functions of RMPCT

Profit Controller's patented technology generates controller applications that are inherently more robust and offer a number of innovative features that yield better control performance and higher application uptime with less application maintenance.

16.4.3.1 Special Feature to Handle Model Error
Unavoidably, however, all models contain some error. One of the most important aspects of a multivariable controller is its ability to cope with this error.

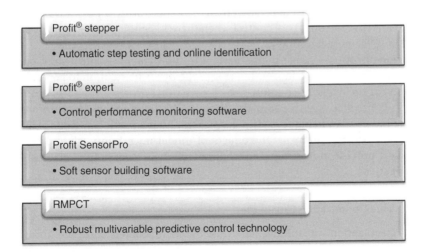

Profit® stepper
• Automatic step testing and online identification

Profit® expert
• Control performance monitoring software

Profit SensorPro
• Soft sensor building software

RMPCT
• Robust multivariable predictive control technology

Figure 16.7 RMPCT product package

The problems caused by model error are more pronounced for processes that have large interactions between CVs that are assigned set points or are at constraints most of the time.

16.4.3.2 Coping with Model Error

RMPCT excels in its ability to provide good control even with large errors in the process model and large interactions between CVs.

Here is a broad, first look at how RMPCT manages this:

- *Limit funnels:* The controller imposes its own constraints on CVs to avoid introducing transient errors.
- *Range control:* You have the option of controlling CVs within a range rather than to set point. This gives the controller greater freedom to reject disturbances.

16.4.3.3 Funnels

Its innovative feature of managing dynamic control moves through *funnels* rather than via specified trajectories provides the controller with additional degrees of freedom to enhance dynamic process optimization.

The shape of the high and low constraint lines resembles a funnel turned on its side. The mouth is opened beyond the operator-entered limits or the CV value at the current time, and narrows to the limits at the horizon.

The controller is obliged to keep the CV within the constraints defined by the funnel, but is allowed to follow any trajectory within these constraints.

The controller automatically determines the amount that the funnel opens beyond the limit or CV value at the current time. This gives flexibility to the control response (such as when handling inverse response dynamics), and still prevents an excessive amount of transient coupling of CV values.

16.4.3.4 Range Control Algorithm

Range Control Algorithm (RCA) has a key profit controller advantage. In many applications, many of the CVs have range limits rather than set points. RMPCT takes advantage of range limits to improve the quality of control when there is significant model error or when a number of CVs interact strongly.

Basically, range control uses range limits directly in the dynamic control solution, rather than using a steady-state optimization prior to the dynamic solution to generate internal set points for the dynamic solution.

This gives the controller the freedom to allow a CV to be anywhere within a prescribed range rather than being forced to keep each CV at a particular value. This patented control algorithm minimizes the effects of model uncertainty while determining the smallest process moves required to simultaneously meet both control and optimization objectives.

16.4.4 Product Value Optimization Capabilities

Product value optimization capabilities are directly incorporated into the Profit Controller. The user can configure certain variables to be maximized or minimized, and can also specify desired process values as targets that will be honored under optimization conditions. The most powerful optimization scenario, however, is the ability to enter process economics directly into the controller in terms of both manipulated and controlled variable factors that support optimization of the overall economics of the process.

This last option is commonly referred to as product value optimization (PVO) because it allows the controller to automatically determine the best economic operating condition of the unit based on input variables such as product prices, feed prices, and utility costs.

16.4.5 "One-Knob" Tuning

One-knob tuning simplifies the engineering effort required to implement and maintain controllers. A single performance ratio is available for each controlled variable to adjust the desired control response independently from the other controlled variables.

In addition to simplified controller tuning, Profit Controller also employs a patented technique that allows the controlled response of feedforward inputs to be tuned faster than the controlled response of feedback inputs. This allows aggressive feedforward disturbance rejection without the introduction of instabilities in controller feedback.

Model flexibility in the overall controller structure allows control parameters, including gains and constraints, to be adjusted while the controller is online. This powerful functionality allows easy implementation of advanced control strategies to optimally solve difficult control problems, such as changes in feed quality, changes in operating mode, process nonlinearity, and the incorporation of rigorous models.

16.5 SMOC–Shell Global Solution

16.5.1 Evolution of Advance Process Control in Shell

16.5.1.1 1975–1998: The Beginnings

The basic concept for APC followed quickly behind the substantial drop in costs in computing power in the 1970s.

Shell was considered as one of the pioneer in MPC development when two groups, one in France and other in the United States, were engaged in the 1970s to develop MPC algorithm for refineries (see Figure 16.8). Later on, these became SMOC and DMS, respectively.

In the late 1980s, engineers at Shell Research in France developed the Shell Multivariable Optimizing Controller (SMOC) (Marquis & Broustail, 1998; Yousfi & Tournier, 1991), which they described as a bridge between state-space and MPC algorithms.

From then, up to 1998, Shell continued to refine MPC algorithm and simultaneously implemented MPC in its own sites and JV companies. Due to implementation in their own sites, Shell had obvious advantages to learn fast from its mistake and make quick roadmaps to develop a more efficient MPC algorithm. Skilled workforce, expertise in MPC project execution, and making the MPC profit in their own sites are the major advantages of Shell MPC team. In this era, strategically Shell confines the SMOC application in its own sites.

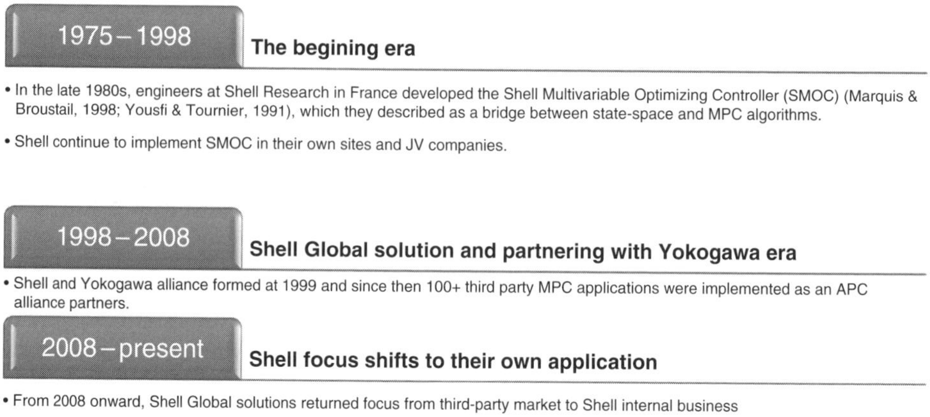

| 1975–1998 | **The begining era** |

- In the late 1980s, engineers at Shell Research in France developed the Shell Multivariable Optimizing Controller (SMOC) (Marquis & Broustail, 1998; Yousfi & Tournier, 1991), which they described as a bridge between state-space and MPC algorithms.
- Shell continue to implement SMOC in their own sites and JV companies.

| 1998–2008 | **Shell Global solution and partnering with Yokogawa era** |

- Shell and Yokogawa alliance formed at 1999 and since then 100+ third party MPC applications were implemented as an APC alliance partners.

| 2008–present | **Shell focus shifts to their own application** |

- From 2008 onward, Shell Global solutions returned focus from third-party market to Shell internal business

Figure 16.8 History of SMOC

16.5.1.2 1998–2008: Shell Global Solution and Partnering with Yokogawa Era

In 1998, Shell engineering staff were put into Shell Global solutions. Shell took strategic decision to take SMOC in third-party market outside Shell assets and Shell JVs. Shell strategically wants to emerge SMOC as a generic MPC product and leverage its strength of large application knowledge of MPC implementation in refining and chemical processes along with best in class algorithm of their SMOC. It chose Yokagawa as a strategic business partner to complement its strength in MPC business. Yokogawa, being a DCS supplier, had obvious advantages of expertise in plant automation and system integrator with IT development skill.

The Shell and Yokogawa alliance formed at 1999, and since then, 100+ third-party MPC applications have been implemented as an APC alliance partners. This alliance was developed based on Shell's long experience with advanced control technology and Yokogawa's extensive global track record in providing superior control system solutions, and is expected to help third party customers achieve dramatic improvements in productivity. Shell focused on Shell and its JVs companies and Yokogawa focused on non–Shell third-party companies. The major achievement of Shell/Yokogawa alliance in MPC field is to get the recognition of SMOC among new customers like Armco, Fuji oil, PTT, and so on. The commercial success of the alliance gives access to financial income for further software development.

16.5.1.3 2008 Onward: Shell Returns to Its Own Application

From 2008 onward, Shell Global solutions returned focus from third-party market to Shell internal business. Shell agreed to continue to support Yokogawa for winning any third-party work and APC alliance with Yokogawa renewed for another few years. However, strategically, Shell has shifted its focus on its own business and Shell JVs sites. From 2008, Shell has continued to upgrade and review its existing MPC application in its own sites. Brown (1999) reported implementation of several large plant-wide MPC projects and major upgrades at LNG sites.

However, in stiff competition in MPC market, both Yokogawa and Shell agreed to start developing the next-generation APC technology. They jointly developed next-generation platform for advanced control and estimation in June 2015.

16.5.2 Shell MPC Product Package and Its Special Features

SMOC is Shell Global Solutions' Multivariable Optimization and Control suite of software packages. SMOC provides the tools necessary to design, implement, and maintain multivariable Advanced Control strategies to effectively improve plant stability and maximize plant profitability for the hydrocarbon processing and chemicals industries. SMOC contains all the common features described above. Special features are highlighted here:

16.5.2.1 Key Characteristics of SMOC
- Highest uptimes in industry (i.e., highest benefits), as claimed by Shell
- Use of unmeasured disturbance models and grey-box models to include apriori process know-how, resulting in high robustness
- Easy-to-use design and simulation kit (offline)
- Embedding in DCS (no special interface software or doubling of databases)
- Shell Global Solutions Advanced Control engineers, with the support of in-house instrumentation and process experts, offer the whole spectrum of services for successfully implementing SMOC, including benefits study, control system engineering, commissioning, and applications maintenance.

16.5.2.2 Applications
SMOC has been successfully applied to over 430+ applications (dynamic figure) worldwide on process units such as crude distillation, fluid catalytic cracking, hydrocracking, lube oil, styrene, ethylene oxide /ethylene glycol plants, and other major refinery and petrochemicals units. SMOC is recommended to safely push a unit toward its constraints, maintain key operating variables at desired targets, while maximizing the unit profit.

16.5.3 SMOC Integrated Software Modules

SMOC involves the following integrated software modules, as shown in Figure 16.9.

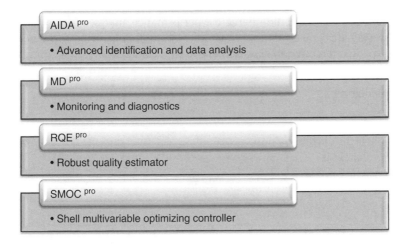

Figure 16.9 SMOC product package

16.5.3.1 AIDA^{Pro} Offline Modeling Package

AIDA^{Pro} is the Shell Global Solutions offline package for the estimation of linear dynamic process models required to implement a model predictive control application, such as SMOC^{Pro}. It makes the intricate science of model identification available in an easy-to-use package and saves valuable project implementation time.

AIDA (Advanced Identification and Data Analysis) is a Windows-based package used to process dynamic plant test data and derive the model needed by SMOC. User-friendly help menus, graphics, and calculated statistics allow the model to be quickly built and validated. AIDA has been designed and tested to be robust in handling noise and disturbances typically present in actual plant test data sets.

Special Features of AIDA^{Pro}

- Advanced Identification Technology—AIDA^{Pro} is computationally fast. Its robust computation engine can process large data sets with unmeasured disturbance effects, and can identify models from both open- and closed-loop data.
- Grey-box modeling
- The knowledge of cause–effect relationships can be incorporated during the process of model identification (grey-box modeling)
- Unmeasured disturbance modeling

Rejecting unmeasured disturbances is one of the main functions of closed-loop control. Unmeasured disturbances include measurement noise, sensor drift, changing ambient identification, and some controller parameters (e.g.,, gains), are incorrectly estimated. This could result in overly aggressive control action. AIDA^{Pro} allows factoring in unmeasured disturbance effects during model identification.

16.5.3.2 MD^{Pro}

MD^{Pro} is the Shell Global Solutions software suite for monitoring all base-level and commercially available multivariable control systems. It also features offline applications that can be used to perform advanced diagnostics on all single output and multivariable control loops, and another for analyzing PID loops only. MD*Pro* technology is being used extensively in many refineries and chemical sites around the world.

Special features of MD^{Pro} include the interactive performance analysis. The offline diagnosis tools use statistical methods to extract the essential control performance information and provide interactive analysis screens to find the root causes of control problems.

- Profit variable calculation
- Benefit forecasting
- Response pilots
- Controlled variable error decomposition
- Degrees of freedom analysis
- Constraint analysis
- PID tuning function

16.5.3.3 RQE^{Pro}

Robust Quality Estimator is the Shell Global Solutions technology for design and implementation of inferred measurements for process monitoring and control. An RQE^{Pro}

inferred measurement is typically calculated using selected process variable such as temperature, pressure, and flow. The RQE^{Pro} technology is used extensively in many refineries and chemical sites around the world.

RQE Offline Package The RQE offline package is a fully Microsoft Windows-based graphical package for fast-track design and maintenance of inferred measurements. Special features include use of Kalman filter to update measurements.

The Kalman filter update option provides superior performance in terms of measurement noise filtering. It also allows the full model to adapt to changing process conditions, such as varying gains due to nonlinear or time-varying processes (e.g., catalyst deactivation or equipment fouling). Compared to conventional bias update mechanisms, Kalman filtering provides a smoother prediction, has proven to give significant additional accuracy, improves the reliability of the online inferred measurement, and reduces maintenance requirements. The offline RQE package generates a model file, including all parameters required for the online implementation of the inferred measurement.

RQE Online Package RQE is one of the modules in Shell Global Solutions' COAST (Control Applications standards) library of online advanced control tools. RQE includes the full range of flexible and configurable tools required for the robust and reliable online implementation of inferred measurements.

- RQE^{Pro} Simulation tool (RQESim^{Pro})
- RQESim^{Pro} is an intuitive, user-friendly offline simulation package for testing the performance of RQE^{Pro} applications. It provides process control engineers the ability to tune RQE^{Pro} applications created, using the RQE^{Pro} offline package.
- RQESim^{Pro} provides troubleshooting tool for existing RQE^{Pro} applications. When used as a maintenance tool, RQESim^{Pro} can monitor model update frequency, prediction errors, and quality measurement indicator (QMI) spike filtering and other tuning parameters in the application.

More than 500 inferred measurements are in operation at Shell Global Solutions' advised sites, in refining, chemicals, and gas-processing applications. With this wide basis of applications, Shell Global Solutions has the practical experience to help its clients select, design, and implement successful inferred property calculations for control, monitoring, and information.

16.5.3.4 SMOC^{Pro}

SMOC^{Pro} is the Shell Global Solutions software package for the implementation of multivariable optimizing control strategies. SMOC^{Pro} provides all the tools necessary to design, implement, and maintain advanced multivariable controllers. This is to improve your plant stability and maximize plant profitability for the hydrocarbon processing and chemical industries.

Applications SMOC^{Pro} has been successfully applied in over 800+ projects worldwide (dynamic figure) on crude distillation, fluidized catalytic cracking, hydrocracking, lube oil, styrene, ethylene oxide/ethylene glycol plants, and other major refinery and petrochemicals units. SMOC^{Pro} can be used to safely push a unit toward its constraints, maintain key operating variables at desired targets, and maximize the profit function with all available operating handles.

Summary of SMOC^Pro Features

- Parametric models
- Grey-box modeling (via intermediate variables)
- Robust unmeasured disturbance model
- Powerful scenario-based simulation environment
- Optimization via bilinear QP or external targets
- Dynamic constraint handling
- Easy to use graphical model builder
- Updatable models for use in blending applications

Advanced SMOC^Pro Features

- Advanced modeling capabilities including intermediate variables, cascade correction of manipulated variables, and explicit options to model the magnitude and dynamics of unmeasured disturbances.
- The concept of sub-controllers allows the control designer to break a unit-wide controller down into smaller sections that ease the tuning of the controller as well as reducing the computing time of the unit-wide controller.
- Robust (bi) linear steady-state optimization allows the control designer to define (bi) linear economic functions. The steady-state optimization has built-in robustification, in order to avoid LP chatter.
- Online updateable model gains allow the control designer to employ gain scheduling and nonlinear control techniques.

Two very useful features that follow from this approach are the possibility to use *intermediate variables* and the use of *cascade correction* for controller model design. The concept of intermediate variables allows the engineer to include more process knowledge in the controller design and to build more robust and transparent controllers. This is also known as the grey-box model approach (as opposed to the black-box approach).

The concept of cascade correction allows the engineer to explicitly model the behavior of base-layer PID controller into the MPC design. This allows for a much more relaxed APC solution, when the base-layer PID controller is capable of rejecting the disturbance.

Originally, SMOC controllers used intermediate variables and cascade correction for relatively small controllers (unit size, e.g., a column or a single cracking furnace). Since the release of SMOC^Pro, it is possible to use the grey-box approach not only for controllers on a unit level but also to build large controllers covering a whole plant. For ethylene plants, this means that a plant-wide controller can be built using intermediate variables to link the unit sub-controllers. In the black-box model, there are inputs and outputs and direct models (transfer functions) between the inputs and outputs. There are no intermediate variables and there is no specific knowledge about the process configuration contained in the model. In the grey-box model, the inputs and outputs are linked via intermediate variables, which represent measured variables (such as intermediate flows or temperatures). In this way, the model represents the actual process much better and allows a more robust feedback mechanism, by using the process measurements for these intermediate variables. It is also more intuitive for engineers and operators to build a multivariable controller this way.

PC-SMOC *Pro **Offline Controller Design Package*** PC-SMOC is a Windows-based package, integrated with AIDA, to design, test, and build SMOC controllers. The process dynamic model is generated by AIDA. Alternatively, the model may be built using the graphical model builder. PC-SMOC allows the user to simulate the controller behavior to test the tuning, robustness to model errors and optimization performance. The final stage in the offline design is the creation of the SMOC controller file, for use by the online controller.

SMOC*Pro* offline package features a powerful simulation environment to test the controller behavior. Scenario-based simulations can be used to test tuning, robustness to model errors and optimization performance.

The online controller incorporates all functions required for the operation of the controller in real time.

16.5.4 SMOC Claim of Superior Distinctive Features

16.5.4.1 Integrated Dynamic Modeling Tools and Automatic Step Tests

It has been claimed that integrated dynamic modeling tools and automatic step tests can dramatically reduce modeling time. Automatic step testing is designed to excite the plant for full dynamic response while continuing process control and economic optimization. The powerful identification and modeling tools allow engineers to quickly extract the dynamic model from the plant responses. This integrated tool can reduce the step testing and modeling efforts by half, as compared with traditional manual step testing.

16.5.4.2 State-of-the-Art Online Commissioning Tools

Shell claims that it can achieve high control performance in less commissioning time with state-of-the-art online commissioning tools. The capability to run a "staged" (read only) version in parallel with the "live" (read and write) version allows the control engineer to validate the modification before going live, checking the controller response with different tuning parameters, changes in controlled variables and manipulated variables, and swap "live" version for the "staged" version, with no down time.

16.5.4.3 Online Tuning

Online tuning supports the control engineer in the validation of integrated estimator model and controller tuning using real process data in real-time, before putting the application in live-active control.

Flexible controller tuning, combined with comprehensive simulation, allows for engineers to test the entire runtime application under different situations, in parallel.

Introducing the new concept of best performance value (BPV): Defined as the most desired point CV operation in the case of infeasibility, the controller optimizes MV move plans to slow the movement of CVs from their BPVs, yielding a superior transient dynamic performance of the controller.

16.5.4.4 Advance Regulatory Controls

Brown (1999) said that Shell standard Advance Regulatory Control modules can be deployed rapidly to achieve stable control prior to introduction of Advanced Process Control. Advance Regulatory Controls are field-proven, user-friendly, and off-the-shelf standard function libraries designed to stabilize base layer control and can be used in conjunction with multivariable optimizing control and RQE to achieve desired control and operating objectives.

- *The Shell Surge Volume Control (SSVC):* SSVC is designed to take full advantage of the Surge Capacity available in the plant to achieve a more stable operation. The SSVC module managers the surge vessel's level within specified limits while minimizing flow fluctuation entering or leaving the surge vessel. The algorithm is designed to work not only for one surge volume but also for cascading surge volume such as cascading Distillation Columns. The algorithm will take advantage of periods when there are no large disturbances to bring the level to ideal set point, to make capacity to absorb the next large disturbances.
- *Furnace pass balancing (BALANCE):* BALANCE is designed to evenly distribute the heat energy absorbed by each pass in a multipass heating furnace. The BALANCE module improves furnace efficiency by passing more flow through those passes with higher heat recovery, balancing the outlet temperature of each pass. Each BALANCE module can manage up to two furnaces with up to two cells/chambers. Each cell can have up to 16 passes (coils), and each pass can have up to a maximum of five skin temperatures reading. The BALANCE module also manages the total flow through the furnaces, which is a very valuable tool for the management of heating furnaces.
- *Measurement validation and comparison (MVC):* MVC algorithm is designed to validate two field measurements of the same process output variable for comparison. Deviations between two field measurements of the same process output variable for comparison. Deviations between two field measurements above a reference value are continuously accumulated, and when the accumulated value exceeds a certain limit, an alarm results.

16.5.4.5 Features of New Product

Yokogawa and Shell jointly developed a next-generation platform for advanced control and estimation in June 2015. The major features of this software suite are as follows:

1) *Development of rapidly deployable applications:* With the identification of dynamic characteristic response models, plant step response tests are typically executed over long periods of time to determine how the manipulation of devices affect process values. This software suite carries out this work automatically and displays the results as a model robustness rating. Users can then easily choose the optimum model, greatly reducing the amount of time required for engineering. Models can be recreated easily at any time, which helps maintain advanced control applications in their best state and thus minimizes maintenance work.
2) *Excellent operability and maintainability:* While the monitoring and operation screens for conventional advanced control applications must be created from scratch for a production control system, a hierarchical set of screens is provided as a standard feature with this software suite, simplifying engineering and dramatically

improving operability. These include screens for overseeing the entire system, monitoring target process units, and checking controlled variables and KPIs.

In its displays, this software suite has status marks that indicate whether a process is in a healthy or unhealthy state, so operators can know its status at a glance. These marks indicate the result of changes in operating conditions or changes caused by the deterioration of aging facilities. When such changes occur, the models can be recreated to restore productivity.

3) *Integrated platform for improved engineering efficiency:* When developing an advanced control application, it is necessary to build modules for multivariable model predictive control, soft sensing, the interface with the production control system, calculation, and other functions, and then combine them. With this software suite, these modules can be built and operated from the same platform, improving engineering efficiency.

16.6 Conclusion

Over the years, MPC has decisively demonstrated its value as a best practice by increasing throughput and improving yield, energy usage, raw material usage, product quality, safety, and responsiveness. After completing its initial building phase (1970–1990), from 1990 onward MPC established itself as a proven technology that reduces process variability and inefficiency, improves product consistency, increases throughput by allowing operations to push constraints to the limits and achieve higher return on assets. Currently, MPC vendors are coming up with new advances of technology that considerably reduce the implementation timing, have sophisticated modeling techniques and easy way for online tuning. However, the benefits of MPC application gradually decline over the years if not properly maintained. So, postimplementation monitoring and proactive maintenance is the key to sustain the benefits. There is a scarcity of MPC engineers who really understand the technology well. Most of the MPC engineers are busy implementing the application, so the postimplementation monitoring gets ignored.

For plants that have already implemented the MPC, the best solution to tackle this issue is to develop in-house MPC experts who can monitor and perform proactive maintenance to sustain the benefit. Some plants without proper understanding of the technology starts to implement real-time optimizer (RTO) above ill-configured MPC, which actually make the situations more complicated. Before attempting RTO, plants should make an all-out effort to run the base layer PID control and MPC efficiently first. Then add RTO if required.

Technology-wise, MPC has matured enough, but it has long way to go in its implementation style. Experts are trying to develop more sophisticated step testing methodology, online adaptive model building, and other processes to make its implementation more effective. However, there is no uniformity in implementation style across industry among various vendors. MPC performance still depends on the expertise of its implementation team. Effective procedures and uniform implementation strategy must be developed in the coming years to make the performance of MPC independent of efficiency of the implementation team. Standards (like API, TEMA, and ASME) of MPC technology are still a dream—they must still be developed. All the details of best practices in MPC implementation need to be published and made available to common practitioners.

References

Bitmead, R. R., Gevers, M., & Wertz, V. (1990). Adaptive optimal control: The thinking man's GPC. New York: Prentice Hall.

Bowen, G. X. D. F. X. (2008). Application of Advanced Process Control, RMPCT of Honeywell, in DCU. *Process Automation Instrumentation*, 4, 013.

Brown, A. (1999). Yokogawa, Shell form advanced process control alliance. *Hydrocarbon Online.* https://www.hydrocarbononline.com/doc/yokogawa-shell-form-advanced-process-control-0002.

Cutler, C. R., & Ramaker, B. L. (1979). Dynamic matrix control—A computer control algorithm. AICHE national meeting, Houston, TX, April.

Cutler, C. R., & Ramaker, B. L. (1980). Dynamic matrix control—A computer control algorithm. In Proceedings of the joint automatic control conference.

Cutler, C., Morshedi, A., & Haydel, J. (1983). An industrial perspective on advanced control. AICHE annual meeting, Washington, DC.

DMC Corp. [DMC] (1994). Technology overview. Product literature from DMC Corp., July 1994.

Garcia, C. E., & Morshedi, A. M. (1986). Quadratic programming solution of dynamic matrix control (QDMC). *Chemical Engineering Communications*, 46, 73–87.

Honeywell Inc. (1995). RMPCT concepts reference. Product literature from Honeywell, Inc., October 1995.

Lu, J., & Escarcega, J. (1997). RMPCT: Robust MPC Technology Simplifies APC. In AIChE Spring Meeting, Houston, TX.

Lundström, P., Lee, J. H., Morari, M., & Skogestad, S. (1995). Limitations of dynamic matrix control. *Computers & Chemical Engineering*, 19(4), 409–421.

MacArthur, J. W. (1996). Rmpct: A new robust approach to multivariable predictive control for the process industries. In The 1996 Control Systems Conference (pp. 53–60).

Muske, K. R., & Rawlings, J. B. (1993). Model predictive control with linear models. *AIChE Journal*, 39(2), 262–287.

Qin, S. J., & Badgwell, T. A. (1997, June). An overview of industrial model predictive control technology. In AIChE Symposium Series (Vol. 93, No. 316, pp. 232–256). New York: American Institute of Chemical Engineers, 1971–2002.

Qin, S. J., & Badgwell, T. A. (2003). A survey of industrial model predictive control technology. *Control Engineering Practice*, 11(7), 733–764.

Ruchika, N. R. (2013). Model predictive control: History and development. *International Journal of Engineering Trends and Technology (IJETT)*, 4(6), 2600–2602.

Yousfi, C., & Tournier, R. (1991). Steady-state optimization inside model predictive control In Proceedings of ACC'91, Boston, MA (pp. 1866–1870).

Index

Multivariable Predictive Control: Applications in Industry, First Edition. Sandip Kumar Lahiri.
© 2017 John Wiley & Sons Ltd. Published 2017 by John Wiley & Sons Ltd.